Construction Program Management – Decision Making and Optimization Techniques

Ali D. Haidar

# Construction Program Management – Decision Making and Optimization Techniques

Springer

Ali D. Haidar
Dar al Riyadh-Engineering and Architecture
Riyadh
Saudi Arabia

ISBN 978-3-319-35648-8 ISBN 978-3-319-20774-2 (eBook)
DOI 10.1007/978-3-319-20774-2

Springer Cham Heidelberg New York Dordrecht London

Softcover re-print of the Hardcover 1st edition 2016


Printed on acid-free paper

Springer International Publishing AG Switzerland is part of Springer Science+Business Media (www.springer.com)

# Preface

The recent trend in the construction industry is how to manage large and complex programs. Multiple projects being built simultaneously, as part of government or private programs, will need structured and sophisticated program management techniques in order to deliver the vast and complex building works at hand with a relatively short period. Without proper program management procedures, these often huge, complex, multi-projects can take decades to construct with draining budgets. With program management, the building works can be relatively short spanning years with significant cost reduction.

Although program management methods have been applied successfully in the USA (NASA and ARMY programs as an example), Europe (Marshall Plan), and some of the United Nation programs, there have not been literature nor research to sustain the theories behind the successful implementation of these methods nor their proper and scientific know-how. All the focus has been on project management techniques with their apparent short-folds for program constructability.

This book will look on the different program management methods, ranging from simple decision-making techniques and statistics analysis to the more complex linear programming, and how program managers, directors, clients, stakeholders, contractors, and consultants can benefit from the availability of these different techniques. The book is unique in a way as it looks on how to apply new and developed techniques to optimize for the delivery of programs mainly in the field of artificial intelligence especially knowledge-based systems and genetic algorithms.

The author's unique experience in complex management, program management, and his past research and studies in analytical analysis and mathematical modeling and artificial intelligence has induced him to write this book to well inform readers about the different techniques that can be applied for future program execution.

No doubt indeed, the future of the construction industry will be in how to execute programs, especially the Middle East war torn areas such as Syria, Libya, Iraq, and now Yemen. As well, African countries and Asian countries will need to be able to execute well-managed programs to build their infrastructure with limited time and constraint budgets.

# Contents

# Chapter 1
# Program Management Perspective

**Abstract** Program management is concerned with the construction of a group of related projects, carried out to achieve a defined objective or benefit to a client, falling under the auspices of a program. The program management process balances the key program constraints and provides a tool for making decisions throughout the program cycle based on benchmark values, performance metrics, established procedures, and the program aims. In order to find better solutions for many aspects of a program, including planning and scheduling, distribution of resources such as labor and staff, optimizing the procurement process and minimizing costs while achieving the program objectives, a program manager must be familiar with the fundamental methodologies of heuristics methods, operations research, and sophisticated intelligent techniques to deal with these complex issues. This book will focus on the fundamental rules in planning and scheduling for program activities, the application of methods and techniques in the decision-making process for often very complex situations, operations research and optimization and mathematical modeling, and lately, the use of complex, efficient, and intelligent systems in order to provide optimal solutions to program managers. The book serves as an introduction for program managers for the above and introduces the basic elements in critical path method (CPM), statistics and forecasting methods, linear programming, knowledge-based systems, and genetic algorithms and their applications as decision-making tools in key areas in program management.

## Introduction to Program Management

Program management applies techniques that allow organizations to run multiple related projects concurrently and obtain significant benefits from them as a collection. A group of related projects, not managed as a program, is likely to run off course and fail to achieve the desired outcome.

Program management is a fairly new technique and as such is not always well understood. It is, however, clear that program management is an area of growing

 
A.D. Haidar, *Construction Program Management – Decision Making and Optimization Techniques*, DOI 10.1007/978-3-319-20774-2_1

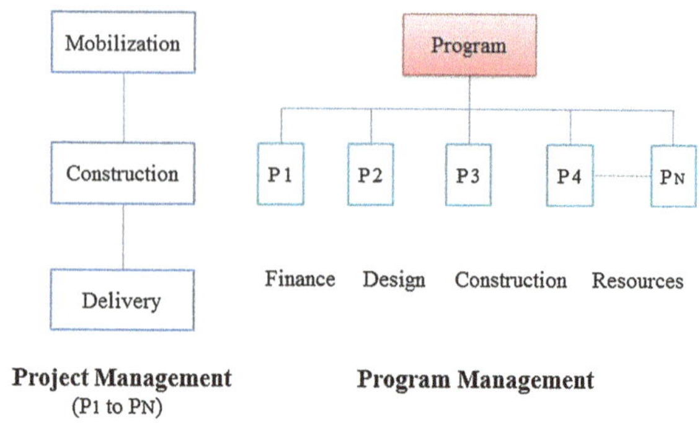


**Fig. 1.1** Difference between project management and program management

interest for organizations involved in many projects related or complimenting each other or a very large project that is split into subprojects in order for it to become manageable.

Traditional project management models and techniques do not recognize the reality of today's organizational structures and workplace priorities, nor do they leverage the potential benefits that accrue from multi-skilled, multi-location teams. Effective program management includes strategies, tactics, and tools for managing the design and construction delivery processes and controlling key factors to ensure that the clients receive facilities that match their expectations as intended to function. Improvements to the processes, when applied directly, contribute directly to reduced operational costs and increased satisfaction for the entire program (Haidar et al. 2014).

Therefore, a program management team can manage the full scope of work and the range of activities necessary to keep complex multi-disciplinary projects ranging from a small number of 2 or 3 projects to a large number running into hundreds of projects, often worth billions of dollars, firmly on time, budget, and the specifications undertaken. Figure 1.1 shows the difference between project management and program management.

One of the reasons that clients choose program management is its ability to provide all the necessary services in-house, eliminating the need for multiple consultants. This capability is changing the facet of management of construction projects, when built more than one at once, as the program manager fully integrated teams offer services and support in design, construction, procurement, project controls, safety, quality, and operations and maintenance, thus eliminating any delays and un-coordination of these activities and hence the optimization of the selection of resources, maximizing the efficiency of the construction, and minimizing costs restraints.

As a general rule, the program phases are subdivided as follows. Figure 1.2 shows the main different phase of program management which can be subdivided as follows:

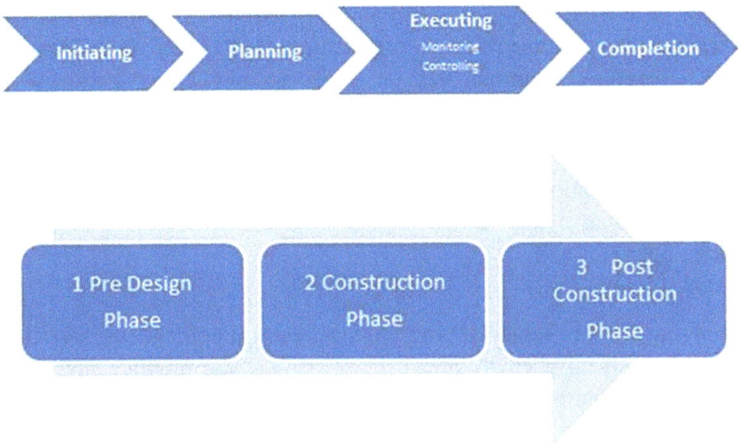


**Fig. 1.2** Different phases of program management

(1) Pre-construction Phase—program managers usually

- work within desired delivery system
- manage planning and design
- evaluate potential sites
- assist in selection of design team
- maximize front-end planning with early feasibility studies to reduce problems during execution
- establish cost and time parameters and prepare bid packages
- offer value engineering input and cost analysis, and
- serve as owner's agent and supplement existing staff.

(2) Construction Phase—program managers usually

- award contracts
- manage construction and coordinate all subcontractor activities
- manage material procurement
- monitor costs and schedules
- maintain quality control
- provide ongoing communications and status reports, and
- monitor contractor's safety

(3) Post-Construction Phase—program managers usually

- develop punch lists
- monitor implementation of punch lists
- resolve outstanding issues
- oversee the systcms and equipment operations training, and
- remain engaged throughout the warranty period.

A group of related projects are an integral part of a program and must be managed within the program parameters to achieve their objectives and the program aims. It is, therefore, important that programs are run within a framework that ensures there is a focus on the overall strategic objectives. The four basic stages in program management are as follows:

- Program identification
- Program planning
- Program delivery; and
- Program closure.

By applying the four stages of program management outlined, organizations will have created an effective environment in which they can monitor and control the progress of their programs, improving the chances of bringing them to a successful conclusion.

These stages take the program from initiation, based on strategy or a desire for change, to the final realization of a defined business objective or benefit (Haidar et al. 2014).

## Program Framework

To leverage maximum benefit from program management, it is important to work within a framework in order to bring project management under control. The framework ensures that there is a focus on delivering the vision or strategy as opposed to the technical delivery of individual projects. The key areas in the framework are as follows:

1. Vision including aims, objectives, and responsibilities. These summarize the high-level strategy or idea to drive the organization toward a goal, benefit, or other desired outcome. The vision will usually be a brief statement of intent communicated down from the management or leadership. It is important that the vision has a high-level sponsorship and commitment for it to be successful. Senior management responsibilities are an integral part of the process which shape the program and project managers' roles.
2. Design and approach. These are the ways in which the projects, which make up the program, are interlinked together. In this process, the program manager considers which projects have dependencies on others and, therefore, which should come first can run concurrently and those that come last.
3. Resourcing. Resourcing looks at the scheduling and allocation of resources. Short-term and longer-term views should be taken. For the projects that will start straightaway, it is important to identify resources and early. For later projects, required resource levels should be identified but decisions are not needed at this stage.

4. Benefits realization. Benefits realization is the process at the end of the program by which the benefits identified at the beginning of the program and measured. It is the responsibility of the program manager to demonstrate to the steering committee that the desired benefits have been realized.

A proper and well-organized framework will provide the following:

- A focus on delivering major organizational changes or benefits
- Greater control through visibility of all projects in the program
- An understanding of projects dependencies
- Clearly defined roles and responsibilities
- A single line of communication to the steering committee or client
- Optimized use of resources across projects
- Ability to leverage economies of scale and maximize value
- Management of risk across related projects, and
- Mechanisms for measuring benefits realization (Watt 2014).

## The Role of the Program Manager

A program manager is often called program director, program leader, project manager, or project director whose main role is often to oversee multiple project managers that are executing various aspects of the program of work. The program manager range of responsibilities varies from the initial feasibility of the program, the establishment of the scope of work from the master plan, to the comprehensive design and eventually the execution of the projects. Figure 1.3 shows a sample diagram of the role of the project manager.

The program manager is, generally, responsible for the delivery of the construction of the program projects through the management of institution

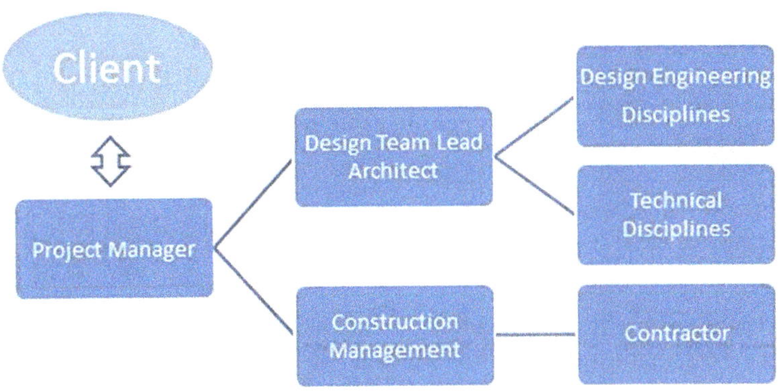


**Fig. 1.3** The role of the project manager

relationships; pre-project planning and feasibility; master planning and completion of design documents including conceptual design and design development; project budgets and schedules; procurement of service specialists providers; administration of contracts; appointment of the contracting team; construction supervision and management; project safety; inspections; furniture, furnishings, and equipment; warranty; and project close-out. In many cases, the program management role extends in the facility management aspect which can lead to the role of the program management team extending over the period of 10 years for the very large programs.

As the clients' single point of contact, the program manager integrates the activities of all designers and architects, contractors and subcontractors, and specialist consultancies such as facility management, quality control, health safety, quantity surveyors, and contract managers, often from diverse backgrounds and cultures, to ensure the success of the overall program.

In summary, the program manager scope of work varies into multi-functions such as:

- The allocations of responsibilities between the various parties involved in the program
- Achieving client objectives
- Identify the contractual matrix of duties between the designers, the main contractors, and the various consultants and subcontractors
- Establish budget control
- Minimize the client's financial burdens
- Impose quality and safety control for the program; and
- Execute the complex construction of each project in the program.

A program manager, hence, has duties more widespread and global than a project manager. He is functioning at a higher level and overseeing many project managers at once. Hence, a program manager usually has long experience of megaprojects, worldwide. Figure 1.4 demonstrates the role of the program manager and how clearly it differs from the role of the project manager illustrated in Fig. 1.3.

The services of a program management team or a program manager include the following:

- Design projects management
- Planning and environmental projects management
- Construction management
- Procurement
- Administrative and financial support
- Human resources and participation
- Information management; and
- All-inclusive projects control.

A program manager can manage programs with full disclosure to and involvement of the owner, as well as full information to the public (as agreed by the owner) and full compliance with all laws, rules, regulations, permits, and requirements. His capabilities aggregate the experience, resources, and expertise in planning,

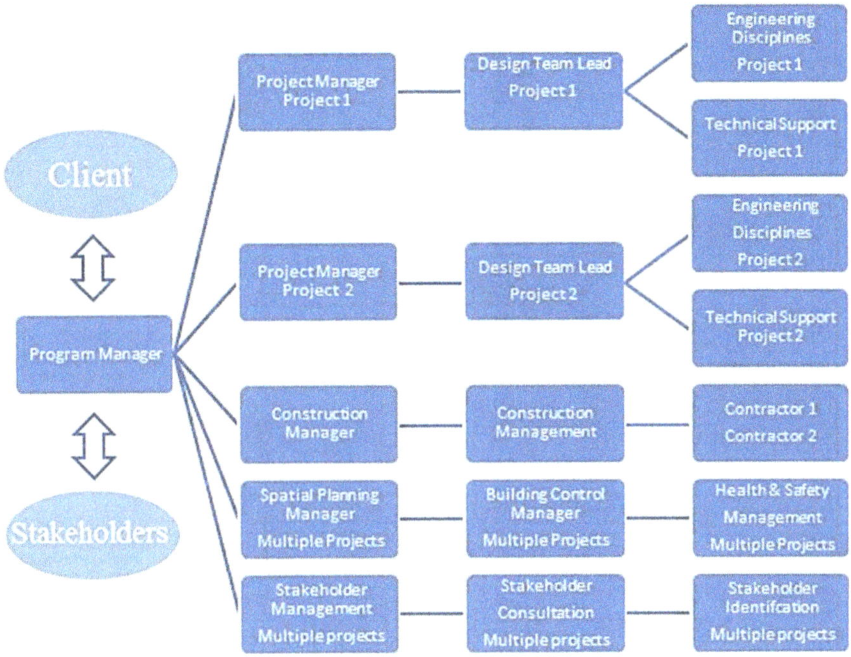


**Fig. 1.4** The role of the program manager

engineering, and management. He can take an entire program from first concept through the master plan and initial design; environmental impact study processes and permits; financing and funding; development design, procurement, and construction; and start-up and operations to ultimately hand over and facility management.

## Program Management Objectives

Program management defines the long-term objectives of the client and the stakeholders (together called the organization herein). Once these long-term objectives are established, the organization identifies programs that help achieve these objectives and thinks carefully about the benefits these programs are designed to bring about. Some of the objectives that can help shaping program management are given below:

- advise that the organization setup assorted structures to manage the program and keep the strategic objectives in mind
- provide the organization a chance to stop and take a look at what has changed, what is to change next and to compare all of that with those highly significant overall objectives

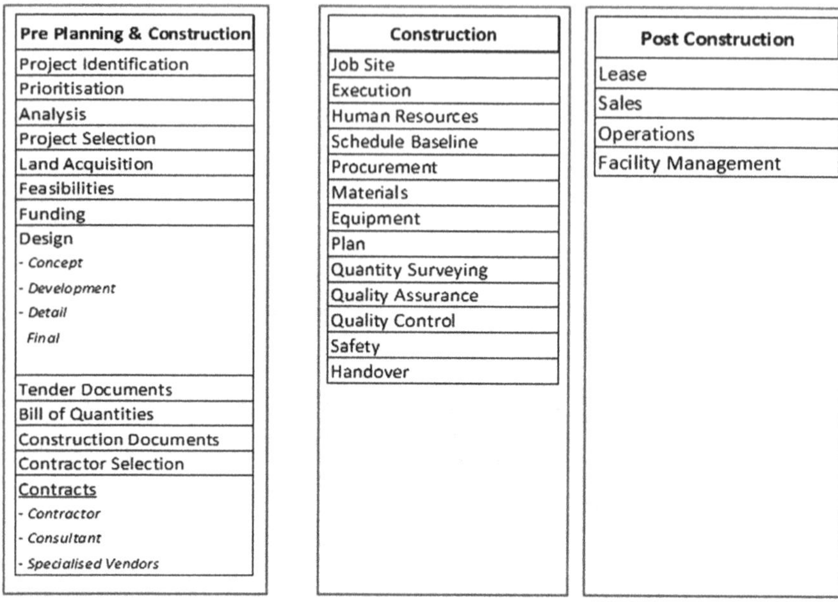


**Fig. 1.5** Program management conceptual framework

- coordinate management of a portfolio of projects to achieve a set of predefined objectives such the use, sale, or lease of the projects. The projects can be houses, buildings, airports, train stations, electrical stations, desalination plants, roads, tunnels, and so on; and
- sequence the planning and monitoring of tasks and resources across a portfolio of projects. Figure 1.5 shows the different components in a typical program management conceptual framework ranging from the preconstruction phase to the construction phase and ultimately to the post-construction phase.

Program management sought to determine whether its objectives can be clearly defined to include a set of knowledge, skills, and abilities, which are unique to achieve its aim and not possessed or practiced by other professions inside or outside the construction industry; whether specific, objective criteria exist (including measures of cost savings or cost avoidance), which decision makers in the public sector could use to determine the feasibility and benefits of using different contractors; and the extent of current regulatory controls governing the use of program management.

This goal is what the multi-project manager, and the projects team, strives to achieve under given specific schedule constraints and a set of predetermined benefits. Although, program management has been established in many sectors of economy, yet its understanding and practice in the construction sector is still in its infancy stage. In program management, issues such as management, organization, financing, optimal resource utilization, and realization of stakeholder needs are basic considerations.

The goals and aims of program management are mainly:

- optimize of the schedule across the program and deliver incremental benefits, as well as enable staffing to be optimized in the context of the overall program's needs
- creation value by improving the management of projects in isolation, especially where the working environment is not only made up of a many of small projects, but also where integration of projects is crucial in terms of development and deliverable for a competitive success
- the use of resources on sharing basis from the common resource pool. This will reduce the amount of resources as needed for the required activities.

The traditional project management approach executes a project as an individual endeavor. Hence, each project has its own resource requirements in the organizational resource pool. However, when all the similar work is executed as a program or multi-projects, the usage of resources will be optimized on sharing basis from the same resource pool. In this way, a rational controlling and planning will also help to reduce the resources usages to a considerable level and support the agenda of sustainability as well.

## Necessity for the Development of Structured Systems

A common perception is that the current system of managing the construction of projects is a failure because facilities may not be completed on time and within budget or may not meet quality requirements. Several management and oversight solutions are available to address such problems and to help insure construction of a quality facility that will meet an entity's needs and be completed on time, using the benefits of sophisticated management tools as will be discussed in this book.

While the goal of any construction program is finished, projects which meet the owner's needs, the program manager and his team have different perspectives and competing interests in the projects. The program manager wants quality-constructed projects which are delivered on time at the lowest possible cost, while the constructing team members want to maximize profits in the course of fulfilling their contractual agreements.

Ultimately, the program manager is responsible for ensuring that the finished projects are within the global budget and meet the needs of the client. To assist them in managing and monitoring construction projects, program managers may enlist assistance of independent third parties. These parties should represent the interests of the program manager and exercise oversight independent of the constructor team.

Program inception and preliminary planning require thoughtful definition of goals and needs (program scope); master planning to accommodate anticipated future needs; evaluation of program alternatives; identification of sites requirements; funding requirements; budgets authorization cycles and/or financial impacts; and program phasing.

The risks associated with making mistakes in this part of the process are great, since their impact will be felt across the program development process and in the final program results. There are tools available that help define the goals and objectives for the program that can optimize the program benefits and minimize the associated risks.

Construction projects are usually classified into residential schemes, commercial buildings, industrial setups, and infrastructure/heavy construction works. In a multi-project environment, these different types of projects share common resource pool for example material, equipment, and manpower. Thus, program management can be of great benefit since:

- it provides an integrated and structured approach in order to align, allocate resources, and execute plans to manage a number of related construction projects to achieve optimum benefits
- it also provides an optimal, structured, and mathematically tangible approach of sharing of common resources to related projects which is a vital aspect of program management
- the optimal allocation and resource leveling
- it helps to reduce idle time as well as assist in identification the project interdependencies and thereby cut down the frequency of work backlogs, rework, and delays
- it has also been observed that the execution of parallel projects in program management supports the knowledge-sharing among projects and hence, save expenditures that may occur regarding extra employment and staff training; and
- Furthermore, it will also facilitate in choosing the economical methods of working by emphasizing reusability of equipment and facilities instead of procuring new ones.

Hence, centralized management of projects via program management is considered to be cost effective and it will support financial optimization in construction industry through minimal and best utilization of resources.

Other examples of deliverables are heavy civil works and infrastructure development projects. However, these projects are unique in nature as each of them has its own characteristic. Nevertheless, all of them require same basic resources and processes for construction works.

Obviously, the important role in future development of program management in the construction industry will be the application of new techniques and methods, which could raise the efficiency of delivery. Traditional optimization, statistical, and econometric analysis approaches used within the engineering context are often based on the assumption that the considered problem is well formulated and decision makers usually consider the existence of a single objective, evaluation criterion, or point of view that underlies the conducted analysis. In such a case, the solution of engineering problems is easy to obtain. But in reality, the modeling of engineering problems is based on a different kind of logic taking into consideration the conflicting aims of decision makers, the existence of multiple criteria, and the complex, subjective, and different nature of the evaluation process. Therefore, multiple criteria methods contribute in engineering context through the identification of the optimal alternatives taking into account the conflicts between the criteria and the revealing the preferences.

Resource optimization is the set of processes and methods to match the available resources (human, machinery, financial, etc.) against the needs of the organization in order to achieve established goals. Optimization consists in achieving desired results within a set time frame and budget with minimum usage of the resources themselves. The need to optimize resources is particularly evident when the organization's demands tend to saturate and/or exceed the resources currently available.

When a program is managed using the philosophy of intelligent management, then resources optimization is strictly linked to the concept of constraint and a systemic vision of the program. Indeed, without a systemic vision of the organization, the program manager is unable to identify the global effectiveness of resource allocation and will run the risk of using resources available mainly to respond to emergencies that daily occur in the various parts of the program.

Intelligent management sees the structure of a program as a network of projects which cut across organization functions, in contrast to the hierarchical view of an organization divided up into departments unable to recognize precise patterns and rules of interdependencies.

An efficient use of resources to carry out a project requires us to:

- Have a shared vision of the global goal to be achieved (remove unnecessary protection from individual tasks)
- Eliminate multitasking (increased effectiveness in the tasks)
- Identify the constraint (the critical chain) and protect it with a buffer of time (thus protecting the project from variation)
- Carefully manage the operational phases of the project (capitalize on time gained); and
- Carry out a statistical analysis and forecasting techniques of the project buffer consumption.

The situation becomes more complex when multiple projects have to be managed, and possibly by different contractors.

Early investment in planning, programming, and designing can help deliver these benefits and avoid unnecessary costs and delays. Contemporary institutions and organizations are increasingly realizing that traditional methods of management—based on the same approach to every project—cannot meet the needs of today's economic, social, and business environment. Additionally, the processes can be streamlined based on technologies and efficiencies not previously available. The responsibility for delivering a program as planned rests with the entire team. When evaluating options, the whole-life value should be considered and not limited to the short-term initial investment. Factors that affect the longer-term costs of a program, such as maintainability, useful service life, and resource consumption, should be integrated into the decision matrix (Tam et al. 2007; Heiman 1987; Aronson and Zionts 2008).

## Program Planning

The planning stage is where the design of the program takes place. The program manager, in order to plan and schedule the projects comprising a program, will in this order:

1. defines clear objectives,
2. agrees an approach,
3. agrees roles and responsibilities with the team,
4. sets up communications channels,
5. agrees priorities of the projects that make up the program, and
6. completes project planning.

It is important at the planning stage to identify adequate levels of resources for the early projects and identify the requirements for later projects.

When dealing with planning a program, the problem becomes much more complex than the planning of a project since the critical path method (CPM) and the networking methods are multi-faceted. Programs that comprise many projects, sometimes as many as few hundred projects running simultaneously and in different geographical locations, need a very well-organized planning method in order to construct the program within the time specified.

Program planning and networking will involve identifying the activities that are fundamentally different than project planning and scheduling as the stages and the interrelationships of the activities between the milestones are unique to each program.

If project planning and networking is almost micro-planning, then program planning and networking is macro where the decision making is done on a higher level and involves activities usually done on a strategic level and includes maximizing cash and financial resources, design stages, mobilization, starting and completion of each project within the program, human resources allocation, heavy equipment distribution, availability and time of delivery of critical materials, the completion of each project within the program, and the facility management if so required.

For a program, although the technique is similar to project management, the variables and the activities are related to the program and the projects are then considered activities of the program.

The essential parts and activities of a program are in sequence:

1. Design,
2. Contract formation and pre-construction studies,

3. Selection of contractor(s),
4. Construction phase of the projects comprising the program,
5. Human resources,
6. Technical support,
7. Supervision, and
8. Facility management.

The most widely used planning and scheduling technique, the CPM, breaks down an entire program into individual activities, estimates feasible durations in which to complete the activities, controls overall program completion and produces numerous activity paths while creating relationships among them. The longest path of the resulting schedule is called the "critical path"; it consists of activities that, if delayed, will extend the program beyond its predetermined completion date. In addition to the critical path, there are other various side paths called non-critical paths. If affected by improper scheduling or performance delays, these paths—not initially planned as critical—could become critical and thus alter the original critical path.

A CPM schedule is designed to advise involved parties about the relative importance of performing certain activities within the program completion parameters. For example, it indicates to participants whether their work is critical, non-critical, or has any float associated with its performance.

Planning of the complex sequences of projects within a program, and their dependencies are of the principle skills of the successful contractor or program manager. If delay allegations are to be shown effectively by the contractor and considered properly by the architect/engineer, it will be found that in most situations that a properly prepared CPM indicating quantity output, physical progress, as well as the passage of time is essential. The CPM facts, when married to law, must persuasively demonstrate the desired and sought for result by virtue of the justice, equity, and fairness of each party position. In this book, the author revises planning techniques in program management, delays calculations, and methodologies as well some legal aspects of the application of the CPM for disruptions and claims (Baldwin and Bordoli 2014).

## Decision-Making Approach for Program Management

A program management decision making consists of a large number of interconnected components each of which may serve a different function but all of which are intended for a common purpose. The degree of achieving the common goal is a measure of the decision making for the program effectiveness. Any decision making for program management has a large number of properties. Only some of these properties are relevant to a particular purpose. The values of these properties constitute the state of the decision making for program management and will be described in detail in this book.

The decision-making approach for a program discourages the program manager from initially presenting a specific problem definition or adapting a particular solution to the problem; instead, the approach emphasizes that the problem environment be defined in broad terms so that a wide variety of needs can be identified that have some relevance to the problem. These needs should reflect the complex relations and conflicts implicit to the problem environment. It covers the comprehensive aspects of the engineering practice and the application of modern decision analysis techniques in the planning, scheduling, selection, distribution, appointing the right personnel, and generally making the right choices.

The basic components of the decision-making approach for program management are as follows:

- State of a decision-making
- Environment of a decision-making
- Hierarchical decision-making for program management. (A decision making for program management is composed of subdecision making of a lower order and is also part of a supra-decision making)
- Decision-making analysis; and
- Decision-making models.

Decision-making approach for program management sharpens the program manager awareness of the objectives of the projects he is designing and planning. It will allow him to make precise forecasts, generate large alternatives, and eventually assist him in making a decision. As well, in many situations, decision-making analysis can provide alternative strategies which can be used.

Decision-making analysis compliments the application of many analytical tools such as network analysis, optimization techniques, statistics and forecasting, and artificial intelligent techniques.

Decision analysis refers to a set of quantitative methods for analyzing decisions that use expected utility as the criterion for identifying the preferred alternative.

Decision analysis provides tools for quantitatively analyzing decisions with uncertainty and/or multiple conflicting objectives, and these tools can be especially useful when there are limited relevant data so that expert judgment plays a significant role in the decision-making process. It provides a systematic qualitative approach to making better decisions, rather than a description of how unaided decisions are made (Haidar et al. 2014).

A general decision-making process can be divided into the following steps:

1. Define the problem;
2. Determine the requirements;
3. Establish goals;
4. Identify alternatives;
5. Define criteria;

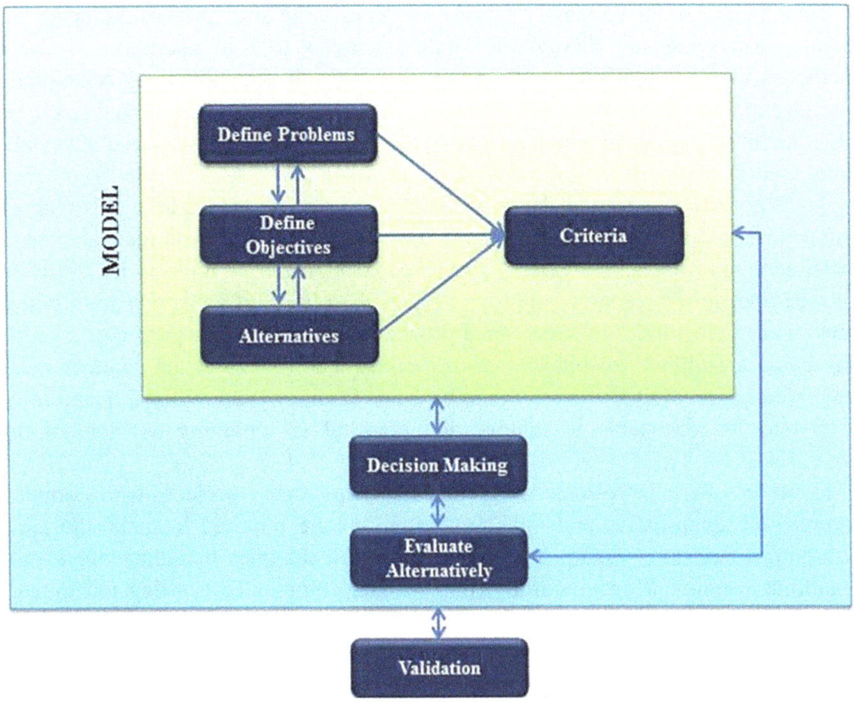


**Fig. 1.6** Decision-making analysis in program management

6. Select a decision-making tool;
7. Evaluate alternatives against criteria; and
8. Validate solutions against problem statement.

The above steps are illustrated through the flow diagram as given in Fig. 1.6.

## Operations Research

Operational research is accepted as a legitimate management tool in defense research establishments and subsequently for efficient resource planning and allocation by many government departments. Business supported the accelerated growth of this discipline by funding real and potential applications. Over period of time, a symbiotic relationship between government, business, and academia ensured the growth and expansion of the discipline for their mutual benefit. During the last 50 years, operational research has evolved as a multi-disciplinary function involving economics, mathematics, statistics, industrial engineering, and management.

As the program management process becomes more structured, more needs for developed systems are thought. Program managers turn to operations research methods as tools to assist them in decision making, selection, planning, scheduling, optimization and basically environments to make their tasks efficient and risk free. This makes operations research as one of the popular managerial decision science tools used in the construction industry.

To be able to become a leader in program management and lead the today highly competitive and demand-driven construction market, pressure is on management to make economic decisions. One of the essential managerial skills is its ability to allocate and utilize resources appropriately in the efforts of achieving the optimal performance efficiently. In some small projects such as the construction of a single unit being a building, hospital, road, or a construction site that the contractor has been building repetitively, the decision based on intuition with minimal quantitative basis may be reasonably acceptable and practical in achieving the goal of the organization.

However, for a large-scale program comprising many projects with complex nature, both quantitative and qualitative analyses are required to make the most economical decisions. Using operations research techniques including linear programming, nonlinear programming, fuzzy logic, complex forecasting techniques, discrete event simulation, and others, program leaders are capable to make high-quality decisions.

Program managers are not expected to be experts in decision science tools and optimization techniques; however, he or she must have fundamental knowledge of such techniques to acquire right resources and to make the most economically sounding decisions for the program as a whole.

Good program management is a combination of optimizing construction cost, high quality execution, distribution of resources, and efficient design. In order to enhance the role of operational research and speed up the process and methodologies of different criteria, project managers should work closely and complement each other's effort. In this process, the program manager should take the lead in the structure, development, and demonstration of sustainable operational research models to optimize the different variables and their interrelationship.

The construction industry should support this initiative and accelerate the transmission of this methodology. This would ensure wealth creation in the short term and sustainable development in the long term. The clients should encourage this initiative by adopting optimized methodologies. Consequently, optimized policy response and its implementation would bring about positive changes in the sociopolitical and economic environment. As a result, sustained use of operational research would be a regular feature in the decision-making process of the government, industry, and society.

Broadly, operational research as a discipline can be classified into three distinct sets of categories. They correspond to tools, models, and methodologies. Tools include ABC analysis, 80:20 rule, and break-even analysis. Blending models, optimized distribution system, portfolio optimization of projects would broadly represent examples under the category of models. Operational research methodologies would include project management systems, multi-criteria optimization, game theory, simulation methodology, data envelopment analysis, enterprise resource planning systems, and conflict resolution methods. The tools, models, and methodologies of operational research have found a variety of applications in different contexts.

For the purpose of this book, linear programming, decision models, network theory, statistics, forecasting methods, knowledge-based systems, and genetic algorithms are presented with detailed examples as these techniques are the most popular in the construction industry and most easily comprehended.

It is imperative to understand that the intent of this book is to present these tools as informative and a guide for program managers. The author has not indulged into the theorem of these techniques as they can be found in many books and references. Chapters 4 and 5 review generally operations research operations methods with specific analysis on statistics, forecasting, and optimization techniques in the form of linear programming (Heiman 1987).

## Significance of Operations Research

The operations research approach is particularly useful in balancing conflicting objectives (goals or interests) where there are many alternative courses of action available to the decision makers. In a theoretical sense, the optimum decision must be one that is best for the organization as a whole it is often called the global optimum. A decision that is best for one or more sections of the organization is usually called suboptimum decision. Operations research attempts to resolve the conflict of interests among various sections of the organization and seeks the optimal solution which may not be acceptable to one department but is in the interest of the organization as a whole. Operations research is concerned with providing the decision maker with decision aids (or rules) derived from:

(i) Total system orientation,
(ii) Scientific methods of investigation, and
(iii) Models of reality, generally based on quantitative measurement and techniques.

Given that operations research represents an integrated framework to help make decisions, it is important to have a clear understanding of this framework so that it can be applied to a generic problem. To achieve this, the so-called operations research approach is now detailed. This approach comprises the following sequential steps:

(1) Orientation,
(2) Problem definition,
(3) Data collection,
(4) Model formulation,
(5) Solution,

(6) Model validation and output analysis, and
(7) Implementation and monitoring.

## Optimization Models for Program Management

One of the main objectives of this book is to establish the importance of optimization techniques approach to the area of construction program management. It provides an approach of optimization through a program of related projects and aligns it with the organizational strategy and objectives.

As stated earlier, the underlying principal of program management is concerned with the management of individual projects when built together under the auspices of a program. Its aim is to balance the specified parameters of cost, time, and quality of multiple projects.

Modeling for decision-making involves two distinct parts, one is the decision maker and the other is the model-builder known as the analyst. The analyst is to assist the decision maker in his/her decision-making process. Therefore, the analyst must be equipped with more than a set of analytical methods.

These are generally mathematical and intelligent representations that describe the interactions between the complex factors of the program management environment and the causal dependencies among these factors so that the analysis can optimize the solutions that may be introduced by large-scale programs.

Generally, the types of models vary and are as follows:

- Iconic (physical presentation)
- Analogue (Schematic)
- Mathematical or analytical
- Computer simulation (Monte Carlo, fuzzy logic,...), and
- Artificial intelligence (knowledge-based systems, genetic algorithms, neural networks,...).

The model-building process includes the following:

- Model formulation
- Model verification (existing data)
- Model application to predict new observations, and
- Model refinement to achieve precision.

The fundamental steps to build the appropriate model are as follows:

- Problem definition and statement of objectives
- Formulation of measures of effectiveness (MOE)
- Generation of alternative solutions
- Evaluation of alternatives
- Selection and implementation, and
- Feedback.

Some of the areas that optimization techniques currently play an important role are as follows:

1. Prioritization of the execution of the projects,
2. Cash flow and financing,
3. Maximizing profit and minimizing expenditure,
4. Scheduling existing resources (production, transportation, cash, personnel, equipment, materials),
5. Acquiring additional resources,
6. Determining what resources are required,
7. Market opportunities,
8. Internal development of financial, human, product, and technological resources,
9. Buying versus leasing and renting, and
10. Time constraints.

We always try to achieve the followings when we attempt to solve complex problems or decisions in a program:

1. Application of a range of mathematical methods;
2. Procedures for selecting the appropriate methods for a specific situation; and
3. Testing the results in a real environment.

Obviously, the important role in future development of the construction industry will be the application of new techniques and methods, which could raise the efficiency of the whole process. Traditional optimization, statistical, and econometric analysis approaches used within the engineering context are often based on the assumption that the considered problem is well formulated and decision makers usually consider the existence of a single objective, evaluation criterion, or point of view that underlies the conducted analysis. In such a case, the solution of engineering problems is easy to obtain. But in reality, the modeling of engineering problems is based on a different kind of logic taking into consideration the conflicting aims of decision makers, the existence of multiple criteria, and the complex, subjective, and different nature of the evaluation process. Therefore, multiple criteria methods contribute to engineering context through the identification of the optimal alternatives taking into account the conflicts between the criteria and the revealing preferences.

The models and techniques that will be reviewed in this book are statistics analysis, linear programming, and artificial intelligent methods, namely knowledge-based systems and genetic algorithms. Those methods are chosen as they are the most understood, researched, and applicable to project and program management. Also, they rely heavily on decision making, knowledge in all its forms and optimization modeling to aid in decision making. The purpose of this book is not to make the reader an expert on all aspects of mathematical optimization, but to provide a broad overview of the field.

## Statistics and Forecasting

Statistics is concerned with scientific methods for collecting, organizing, summarizing, presenting, and analyzing data as well as with drawing valid conclusions and making reasonable decisions on the basis of such analysis.

Statistical methods are less prone to biases and can make efficient use of prior data. Statistical methods are reliable; given the same data, they will produce the same forecast whether the series relates to costs or revenues, to decisions made quantitatively, and to broad number problems faced by the program manager involving resources, procurement, design elements, shop drawings, and quantity surveying. However, statistical procedures are myopic, knowing only about the data that are presented to them. Judgmental forecasting methods are, by their very nature, subjective, and they may involve such qualities as intuition, expert opinion, and experience. They generally lead to forecasts that are based upon qualitative criteria. These methods may be used when no data are available for employing a statistical forecasting method.

However, even when good data are available, some decision makers prefer a judgmental method instead of a formal statistical method. In many other cases, a combination of the two may be used.

The decision on whether to use one of these judgmental forecasting methods should be based on an assessment of whether the individuals who would execute the method have the background needed to make an informed judgment. Another factor is whether the expertise of these individuals or the availability of relevant historical data (or a combination of both) appears to provide a better basis for obtaining a reliable forecast.

In a narrower sense, the term statistics is used to denote the data themselves or numbers derived from the data, such as averages. Thus, we speak of employment statistics, accident statistics, price statistics, and resources demand and availability statistics.

Forecasting is not an exact science but instead consists of a set of statistical tools and techniques that are supported by human judgment and intuition to predict the future (Hastie et al. 2001; Montgomery and Runger 2003).

Forecasting is a very important and strategic task within program management framework. Forecast analysis seeks to identify two of the program management prime objectives which are mainly when will the program be completed and what will it cost. Large variances in costs and schedules will impact the profitability, cost flow, and in extreme cases the viability of projects. A good forecasting technique, therefore, needs to include both the historical trend-based data and competent judgments bases on experience and knowledge.

Chapter 4 reviews statistics and forecasting techniques for program managers.

## Linear Programming

Linear programming is an optimization model developed during Second World War which used to plan expenditures and returns in order to reduce costs to the army and increase losses to the enemy. In operations research, optimization means to find out the maximum profit and minimum loss in any decision-making model which we can apply quantitative techniques for solving its output; thus, we can narrow our choices to the very best when there are virtually immeasurable feasible options. Thus, linear programming is effectively a constrained optimization technique, which optimizes some criteria within some constraints.

Linear programming is a generalization of linear algebra. It is capable of handling a variety of problems, ranging from finding schedules for equipment and transportation models to distributing cements from batch plants to construction sites. The reason for this great versatility is the ease at which constraints can be incorporated into the model and, therefore, linear programming is a powerful technique that is often used by large corporations, and government agencies to analyze complex production, commercial, financial, and other activities when constructing large and complex programs.

In summary, linear programming is a mathematical technique for solving constrained maximization and minimization problems when there are many constraints and the objective function to be optimized, as well as the constraints faced are linear (i.e., can be represented by straight lines). Its acceptance and usefulness have been greatly enhanced by the advent of powerful computers, since the technique often requires vast calculations.

Program managers face many constraints in achieving their goals of profit maximization, cost minimization, or other objectives. For example, in a fixed operational period, a program manager may not be able to hire more labor with some type of specialized skill, obtain more than a specified quantity of some raw material, or purchase some advanced equipment, and he may be bound by contractual agreements to procure a minimum quantity of certain products, to keep labor employed for a minimum number of hours, to abide by some pollution regulations, and so on. To solve such constrained optimization problems, traditional methods break down and linear programming must be used. Linear programming is based on the assumption that the objective function that the organization seeks to optimize (i.e., maximize or minimize), as well as the constraints that it faces, is linear and can be represented graphically by straight lines (Vanderbei 2013).

Since program managers often face a number of constraints and the objective function that they seek to optimize as well as the constraints that they face is often linear over the relevant range of operation, the applications of linear programming can be very useful.

The most difficult aspect of solving a constrained optimization problem by linear programming is to formulate or state the problem in a linear programming format or framework. The actual solution to the problem is then straightforward.

Simple linear programming problems with only a few variables are easily solved graphically or algebraically. More complex problems are invariably solved by the use of computers. It is important, therefore, to know the process by which even the most complex linear programming problems are formulated and solved and how the results are interpreted.

The function to be optimized in linear programming is called the objective function. This usually refers to profit maximization or cost minimization. In linear programming problems, constraints are given by inequalities (called inequality constraints). The reason is that the program manager can often use up to, but not more than, specified quantities of some inputs, or must meet some minimum requirements. In addition, there are non-negativity constraints on the solution to indicate that the program manager cannot employ a negative number of labor or staff or consume a negative number of materials such as steel, cement, timber. The quantities of each product or resources to consume in order to maximize profits or inputs to use to minimize costs are called decision variables.

Standard form of describing a linear programming problem consists of the following three parts:

1. Linear function to be maximized, e.g., maximize,
2. Problem constraints of the following form, and
3. Non-negative variables.

Chapter 5 reviews linear program principles and applications in program management.

## Artificial Intelligent Methods

Where, simple optimization methods for a program, are often linear programming, nonlinear programming or forecasting methods; for a large and complex program, these mathematical methods are not sufficient to deal with the complex mathematics, uncertainty, reasoning, and the risks involved, and therefore, it is often artificial intelligence techniques such as knowledge-based systems, genetic algorithms, neural networks or fuzzy logic are required to solve the problems at hand.

Decision support systems techniques in artificial intelligent techniques are reviewed in Chap. 6 in the forms of knowledge-based systems and genetic algorithms.

Knowledge-based systems are artificial intelligence techniques that embody the knowledge of experts in a specific domain. They are automated reasoning systems which embody useful human knowledge in machine memory in such a way that it can give intelligent advice and can offer explanations and justifications of its decisions on demand.

A typical knowledge-based system has five major components. They are the knowledge based, the working memory, the knowledge acquisition, the inference engine, and the user interface.

A knowledge-based system shell is a knowledge-based system with the specific knowledge taken out. Shells allow the development of knowledge-based system applications easily and quickly (Haidar 1996).

Genetic algorithms are also new techniques that have been emerging from laboratories and are gaining popularity as strong tools for optimization problems too difficult to solve using conventional methods.

The basics of genetic algorithms are as follows:

1. Representing possible solutions to problems as a string of genes on a chromosome,
2. Randomly creating a number (generation) of these chromosomes,
3. Calculating the effectiveness of each chromosomes as a solution to the problem then ranking the chromosomes in order of effectiveness (fitness to survive),
4. Creating a new generation of chromosomes by randomly selecting pairs of chromosomes (parents) and mixing their genes to form child chromosomes. This is done by using crossover, mutation, and adaptation, and
5. Repeating steps 3–4 for a number of cycles.

The randomness of the above process allows the effective exploration of large domains and converges on good solutions relatively quickly.

The performance of genetic algorithms in finding the optimal solution is a function of the implementation of the algorithms process, the size and diversity of the initial population, and the nature of the objectives and constraints of the problem (Haidar 1996).

The first trials of knowledge-based systems and genetic algorithms in engineering domains have proved their potential of being able to solve complex problems which conventional problems are incapable of solving due to the complex symbolic manipulation involved.

The resources selection problem in construction provides an opportunity for the introduction of hybrid knowledge-based systems and genetic algorithms to form intelligent decision support systems. The inherent complexity and the large amount of information needed to build such complex systems require a clear research framework which identifies all the variables that shape the problem and establishing the research model which provide a clear view of how these variables are interconnected to form the proposed system.

The current methods for optimizing the traditional engineering process, such as mathematical modeling and linear programming, have certain limitations. These limitations include the following:

a. Solving a problem with more than one constraint such as time, cost, and locations
b. Solving a problem with more than one dependent variable such as type, number, and life of equipment

c. Solving large problems which can require a great number of equations to correlate the variables, and
d. Oversimplifying the real system in order to make it fit in the mathematical model.

Therefore, the use of artificial intelligent techniques will be fundamental for program managers embarking on optimization modeling where the traditional mathematical methods find inherent problems in solving.

## References

Aronson, J. E., & Zionts, S. (2008, April 28). *Operations research: Methods, models, and applications paperback*. Austin: University of Texas at Austin.

Baldwin, A., & Bordoli, D. (2014, June). *Handbook for construction planning and scheduling*. Hoboken: Wiley. Wiley Online Library.

Haidar, A. D. (1996, February). Ph.D., Thesis entitled *'Equipment selection in opencast mining using a hybrid knowledge base system and genetic algorithms'*. School of Construction, South Bank University, London, UK.

Haidar, A. D., Wells, K., & Thomas, P. (2014, December). *Programme management in construction hardcover*. London: ICE Publishing.

Hastie, T., Tibshirani, R., & Friedman, J. H. (2001). *The elements of statistical learning, data mining, inference and prediction*. New York: Springer series in Statistics.

Heiman, D. W. (1987). *Operations research as applied to construction*. http://pubsonline.informs.org/doi/abs/10.1287/mantech.1.2.20.

Montgomery, D. C., & Runger, G. G. (2003). *Applied statistics and probability for engineers* (3rd ed.). New York: Wiley.

Tam, C. M., Tong T. K. L., & Zhang, H. (2007, January 30). *Decision making and operations research techniques for construction management paperback*. Published City University of Hong Kong Press, June 30, 2007.

Vanderbei, R. J. (2013, July 30). *Linear programming: Foundations and extensions* (4th ed.). Berlin: Springer.

Watt, A. (2014, August 15). *Project management*, BC open text book project, August 15, 2014, licensed under the creative commons attribution 4.0 Unported License.

# Chapter 2
# Decision-Making Principles

**Abstract** In this chapter, the principles of decision making relevant to program management are emphasized in a methodology what is critical to the program manager and conformity on how to standardize the decision-making process. Program management organization theories, structures, and environment will also be analyzed to provide program managers with informative structures and approaches. Large projects and programs are notorious for erosion of value during execution. Decisions made by program managers have a significant impact on the strategic value of the projects delivered and those decisions depend on the information feed on which they are based. This analysis applies theories of organizational behavior, decision making, and other informative tools to investigate the impact of information used by program managers on the strategic value delivered by large programs. This chapter aims to draw attention to how the decision making of program managers during construction execution can impact the long-term strategic goals of programs. Normative and descriptive decision theories and principles, organization theory and structure, chain-in command, systems structures, analysis and environments, formalization, and contingency factors are described in details.

## Decision Theory in Programs

In a general sense, a decision is a position, opinion, or judgment reached after consideration. It is a cognitive phenomenon and the outcome of a complex process of deliberation, which includes an assessment of potential consequences and uncertainties. Decision involves thinking, judgment, and deliberate action to assign irrevocable allocation of resources with the purpose of achieving a desired objective. Basic elements of a decision process include information seeking, ascription of meaning (interpretation), applying decision criteria, and subsequent implementation action.

Decision theory has its root in economic theory, with the assumption that people make decisions to maximize utility on the basis of self-interest and rationality. This, however, does not consider the possibilities or effects of moderating or intervening



A.D. Haidar, *Construction Program Management – Decision Making and Optimization Techniques*, DOI 10.1007/978-3-319-20774-2_2

factors that make decisions reference-dependent. Nonetheless, expected utility theory has been applied in the construction industry with some success and has been the predominant model for normative decision making. The theory is considered idealistic, however, because it focuses on how managers should make decisions rather than how they actually make decisions.

Technical people in the construction industry have been observed to exhibit a tendency for a normative approach to decision making, thereby weakening their ability to deal with uncertainty. Program management is dominated by technical staff and probably more than a few are struggling with tendencies toward this normative thinking phenomenon. An alternative approach is the descriptive decision theory.

Descriptive decision theory deals with how people actually make decisions. It postulates that people make decisions by choosing ways to satisfy their most important needs even if they do not have all the required information and their choice is not optimal. When people are faced with making decisions under uncertainty, they simplify the challenge by relying on heuristics or rules of thumb that are largely rooted in acquired knowledge and past experiences (Dillon 1998).

There are two relevant offshoots of descriptive theory, namely the prospect theory and the theory of bounded rationality. Both theories recognize the ample limitation of human beings to be rational most of the time and postulate that inductive thinking is more natural.

Prospect theory explains decision making under risk, which realistically reflects better the decision processes in megaprojects and programs. The theory distinguishes two phases in the decision process, namely, framing, and valuation. Framing consists of a preliminary analysis of the prospects offered (by the challenge) to the decision maker, leading to a representative construction of his or her perception of the challenge, associated contingencies and possible outcomes. A heuristic simplification of perceived risks or challenges takes place such that the decision maker can make some meaning out of it. During this phase, the quantity, quality, and timeliness of information (information feed) available to the decision maker, together with past experiences and knowledge about relevant subject matter, will have huge effects on how he or she models the possible prospects, which is the outcome of this process. Information timelines have also been hypothesized as a factor due to the time pressure that most program managers are under. Time pressure affects decision making and information suffers degradation when not delivered timely. Valuation follows framing, in which the decision maker assesses the value of each prospect on the basis of an "opportunity–threat" or a "gain–loss" principle and then chooses accordingly. Prospects are consequently labeled, for example, as "opportunity" or "threat." Figure 2.1 shows realization process in program management (Wakker 2010).

Ultimately, the aim of decision making is to minimize uncertainties, which arise from inconsistencies between what actually happens and what was expected to happen. Four reasons, largely related to the management of information to support decisions, have been advanced for why these discrepancies can occur following decisions:

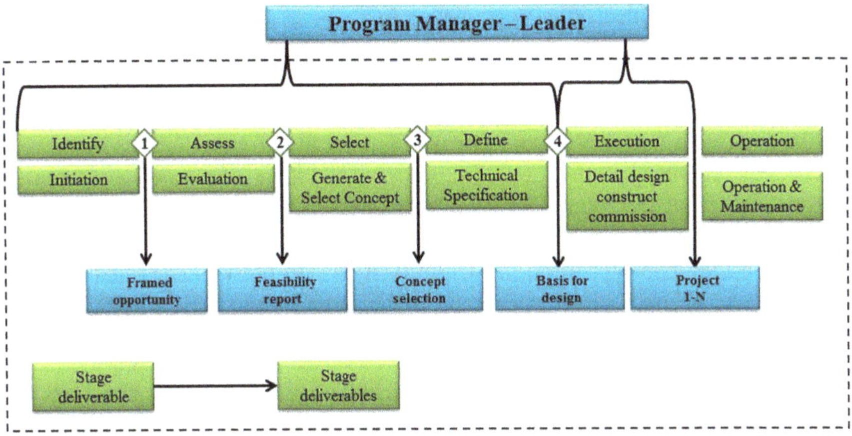


**Fig. 2.1** Basic realization process in program management

1. Mis-information or input data decision process problem.
2. Mis-implementation of what was decided.
3. Change in the assumed context after the decision was made (such as design, resources, or budgetary context around the program).
4. The decision itself may be fundamentally flawed in quality, which would be a problem with the decision approach or process.

## Information Feed in Decisions

The financial and social stakes in programs are so large they can endanger the survival of corporations and threaten the economic stability of some countries they are being built in. Underperformance includes substantial shortfalls in benefits such as financial performance of the delivered projects comprising the program, delays and disruptions and lack of quality in some instances. Therefore, decision making is essential for the main reasons:

1. Program and projects managers' decisions impact the strategic value of assets delivered by megaprojects,
2. These decisions are dependent on the information feed on which they are based,
3. The extent to which managers feel in control influences the scope and quality of information feed,
4. Information feed significantly influences strategic value creation on programs, and
5. Areas of uncertainty may impact long-term success in large programs.

It can be established that the root cause of almost all programs failure can be traced back to human error or misjudgment, and poor judgment can often be traced back to the way the decisions were made. As making decisions is considered the most important job of any executive, the ability to make right decisions on programs should be a principal indicator of professionalism in program management.

Information feed involves searching external and internal environments to identify important issues or events that could affect the program and its objectives. It is a key element of the decision process, enabling managers to formulate expectations about the future. As decision makers will usually have access to far more information than they can deal with, they become selective in favor of information they consider to be most useful. It has been established that decision makers who use more information tend to be more comfortable in dealing with ambiguity and uncertainty and consequently more positive about labeling their challenges.

Program managers who are positive about labeling (as suggested by prospect theory) tend to project positive outcomes with expectations of "gain" or "opportunity" rather than "loss" or "threat." They also tend to have a fair amount of control in organizing or directing the program. In contrast, "threat" labeling implies a negative situation in which a likely loss is projected by the decision maker, and over which he or she feels relatively little control.

Early detection of system disturbances is enhanced through good and timely information feed that allows for pro-activeness. Less timely information is generally considered inferior because the program manager's expectations will contain greater error. On the other hand, decision makers tend to use less information when they believe they are knowledgeable about their business environment or situation than when they feel it is poorly understood. However, decision makers may sometimes not be correct in their judgment. The quality and quantity of information available to decision makers in business organizations has been found to correlate with the quality of their decisions. As program management is similarly underpinned by decisions, one can expect that the information feed to the program manager (as a key decision maker) will influence program performance and derivable strategic value (Eweje et al. 2012).

The extent to which a program manager feels in control of strategic issues is an important influence on how information gathering toward decision support and interpretation will be approached. The level of confidence of being in control would largely be influenced by how the program manager perceives the quality and effectiveness of risk management on the program. The following areas of greatest challenges to mega projects and programs were identified as:

1. Design, including master plan
2. Appointment of consultants
3. Contracting and procurement management
4. Government relations management (the decision mechanisms of host governments are often unclear and can lead to significant complications)
5. Host community relations management
6. Joint venture interface management
7. Health, safety, security, and environmental (HSSE) issues

8. Multi-location management of fabrication and facilities integration
9. Resource allocation
10. Implementation of local content policies
11. Project governance
12. Managing the core program team, including attaining cohesion within the broader team
13. Impact of multi-cultural leadership within the project
14. Facility management (Haidar et al. 2014).

Note that the information feed in support of the program manager's decision will have a significant influence on the level of derivable strategic value. The magnitude of external focus within the information feed in support of the program manager's decisions will correlate positively with the long-term strategic value realized.

## Program Management Quality

Programs are defined as collections of single projects that run concurrently. Fundamentally, these multiple projects must be operated efficiently. However, program management focuses on effectiveness of the execution of the right projects within the program. If a program is regarded as an organization's investment strategy, the right projects would be those that yield the most return on investment for this organization, based on the consideration of a single program and the program level risks.

Thus, program management is a decision-making process that steers the right projects from idea to successful implementation. These decisions are made on present and potential projects and include selection, prioritization, and completion as well as re-allocation of resources across the collection of projects. The process takes the following objectives into account:

1. Information quality is concerned with the availability, comprehensiveness, and transparency of information;
2. Resource allocation quality is related to the speed of assignment, reliability of commitment, and avoidance of conflicts during resource endowment; and
3. Cooperation quality implies empathy and readiness to help project managers and other project teams (cross-project cooperation).

## Organization Theory

Organization theory suggests that the ability of a person within an organization to influence its strategic direction is a function of the amount of resource allocation he or she controls, and not necessarily his or her seniority. The managers of some programs can be responsible for the allocation of between $0.3 and $20 billion for a single program and, therefore, the ability of these senior program managers to influence corporate strategic direction should not be underestimated. Failure of just one program can potentially be disastrous to a contractor or a client.

Some of the topics of particular interest to organization theory are as follows:

1. Goals and value systems,
2. The use of technology and knowledge,
3. The structuring of organizations,
4. Formal and informal relationships,
5. Differentiation and integration of activities,
6. Motivation of program participants,
7. Status and role systems,
8. Organizational politics,
9. Power, authority, and influence in organizations,
10. Managerial processes,
11. Organizational strategy and tactics,
12. Information decision systems,
13. Stability and innovation,
14. Organizations' boundaries and domains,
15. Interface between projects within a program,
16. Planned change and improvement,
17. Performance and productivity,
18. Satisfaction and quality of work life, and
19. Managerial philosophy and organization culture (Haidar et al. 2014).

## Program Organization Structure

Structure may be considered as the established pattern of relationships among the components or parts of the organization to effectively manage and construct a program or manage a portfolio. We consider that the structure of the organization of a program cannot be looked at as completely separate from its functions; however, these are two separate phenomena. Taken together, the concepts of structure and process can be viewed as the static and dynamic features of the programs to be constructed. In some programs, the static aspects (the structure) are the most important for investigation; in others, the dynamic aspects (the processes) are more important.

Static programs relate to schools, buildings, hospitals, airports, roads, and others. Dynamic programs are related to engineering, procurement, and construction (EPC) of power stations, desalination plants, district cooling plants, transportation programs in trains, metros and bus routes, and oil and gas. Renewable energy programs in wind, solar, or water motion are dynamic. Some programs are a hybrid of both static and dynamic. Figure 2.2 shows the generic program organizational structure.

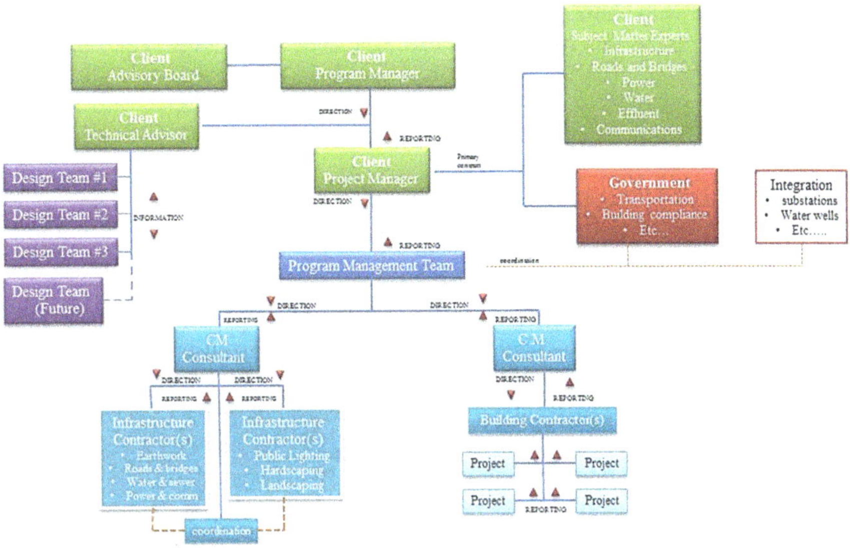


**Fig. 2.2** Generic program organizational structure

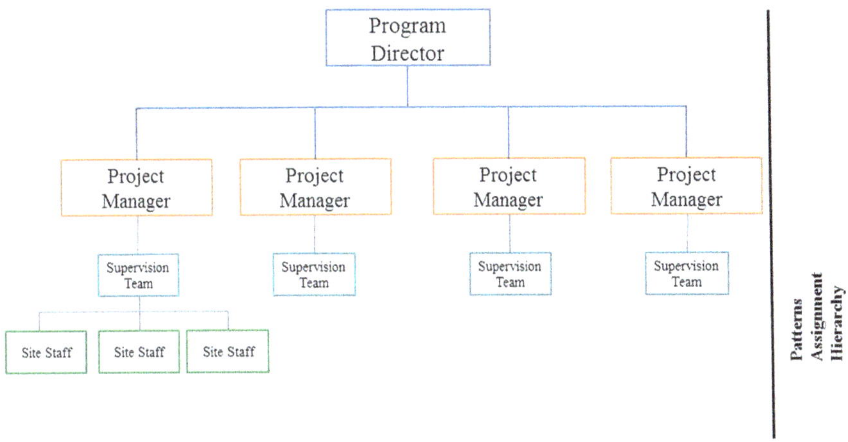


**Fig. 2.3** Formal organization structure

Formal organization is the planned structure and represents the deliberate attempt to establish patterned relationships among projects that will meet the program objectives effectively. Figure 2.3 shows the formal organization structure. The formal organization structure is frequently defined in terms of the following:

1. The pattern of formal relationships and duties. This includes the organization chart plus job descriptions or position guides;
2. The way in which the various projects or tasks are assigned to different departments and/or people in the program organization (differentiation);

3. The way in which these separate projects or tasks are coordinated (integration);
4. The power, status, and hierarchical relationships within the program organization (authority system); and
5. The planned and formalized policies and controls that guide the program in the organization.

The informal organization refers to those aspects of the program that are not planned explicitly but arise spontaneously out of the activities and interactions of the projects. Informal organizations are vital for the effective functioning of the program organization. Informal organization relates to the projects themselves, whereas formal organization relates directly to the upper hierarchy of the program.

It is impossible to understand the nature of a formal program organization without investigating the networks of informal relations and the unofficial norms as well as the formal hierarchy of authority, and the official body of rules. The distinction between the formal and the informal aspects of a program life is only an analytical one and should not be ratified as there is only one actual program organization body (McCullough 2008). Figure 2.4 shows a hybrid formal and informal organization structure.

The concept of a program organization plan implies the process of developing the relationship and creating the structure to accomplish organizational purposes. Structure is, therefore, the result of the planning process. An organization program has a perspective and an action orientation; it is geared toward solving problems and improving performance to construct the projects.

Program organization including planning, orientation, and strategy is never complete; it is a continuing, ongoing process. Hence, a well-designed program is not a final solution to achieve but a developmental process to keep active.

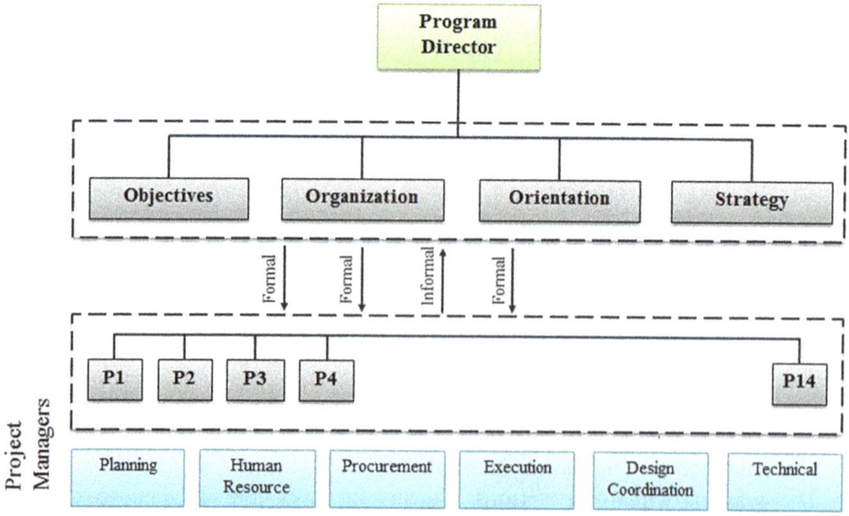


**Fig. 2.4** Hybrid formal and informal organization structure

Coordination of activities within the various projects of a program is an important consideration of the organization structure. Integration is defined as the process of achieving unity of effort among the various sub-systems in the accomplishment of the organization task. The requirements of the environment and the technical system often determine the degree of coordination required. In some organizations, it is possible to separate projects activities in such a way as to minimize their resource requirements.

## Responsibilities and Functions

Structure is directly related to the assignment of responsibility and accountability to various program organizational units. Delegation is fundamental in the assignment of both authority and responsibility. Control systems are based on the delegation of responsibility. Most organizations develop some means to determine the effectiveness and efficiency of the performance of these assigned functions and create control processes to ensure that these responsibilities are carried out.

Traditional management theorists were primarily concerned with the design of efficient decision-making techniques. They emphasized such concepts as objectivity, impersonality, and structural form. The program organization structure is designed for the most efficient allocation and coordination of projects that relate to the different parts of the constructability of the program. The positions in the program structure, not in the people, have the authority and responsibility for getting programs accomplished. Figure 2.5 shows some sophisticated decision-making techniques in construction.

The authority of the program manager is the right to invoke compliance by project managers and staff on the basis of formal position and control over rewards

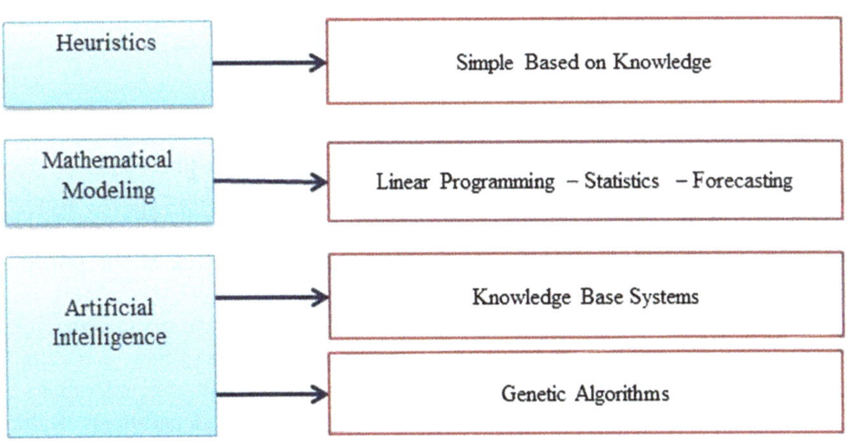


**Fig. 2.5** Decision-making techniques in construction

and sanctions. Authority and responsibility should be directly linked; that is, if a subordinate is responsible for carrying out an activity or a project, he or she should also be given the necessary authority. Accountability is associated with the flow of authority and responsibility, and it is the obligation of the subordinate to carry out his or her responsibility and to exercise authority in terms of the established policies.

This view of authority, responsibility, and accountability provides the framework for much of traditional program management theory.

## The Scalar Principle

There are different factors affecting the chain of command in a program, such as the geographical location of projects, the capability of the engineers, staff and workers, and the similarity of projects comprising the program. Other factors of much importance in constructing a program are the complexity of the projects, the level of the design, the availability of resources, and the technical know-how. In terms of program organization, delegation is very important to keep a tight control on a large number of projects comprising the program.

The program manager must be decisive and authoritarian with respect to the following:

1. Delegate as simply and directly as possible. Give precise instructions;
2. Illustrate how each delegation applies to the program objectives;
3. Develop standards of performance;
4. Clarify expected results;
5. Discuss recurring problems;
6. Seek project managers' ideas about how to construct and manage separate projects and specialist trades such as mechanical, electrical, cladding, and piling;
7. Recognize superior performance;
8. Keep your promises; and
9. Avoid excessive checks on progress.

The scalar principle establishes the hierarchical structure of the organization. It states that authority and responsibility should flow in a direct line vertically from the highest level of the program hierarchy organization to the lowest level. It refers to the vertical division of authority, and responsibility and the assignment of various duties along the scalar chain. Figure 2.6 shows a typical, scalar principle mechanism (Naidu and Krishna Rao 2008).

Although most organization charts are drawn to emphasize the vertical hierarchy and superior–subordinate relationships, very few indicate horizontal interactions, those integrative activities that flow between departments, units, or individuals at approximately the same level, such as the technical different departments dealing with quality assurance/quality control, planning, quantity surveying, cost control, value engineering, procurement, contracts. The function of horizontal relationships

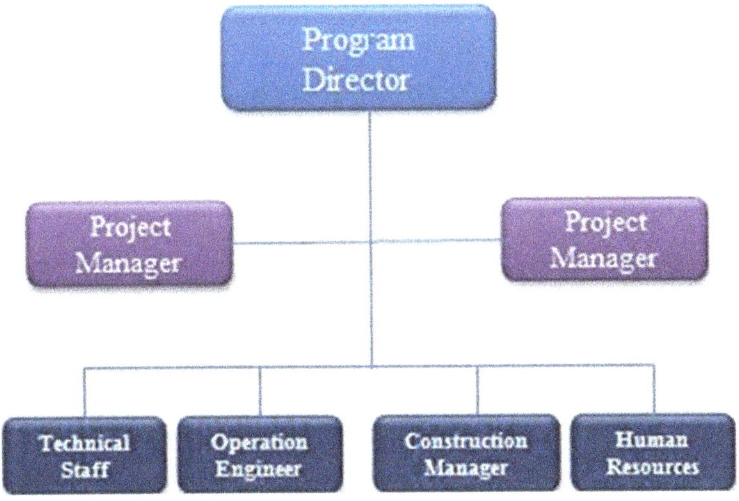


**Fig. 2.6** Scalar principle mechanism

is to facilitate the solution of problems arising from division of responsibilities and the teams working on a program, and their nature and characteristics are determined by the participants having different organizational subobjectives but interdependent activities that need to intermesh.

Figure 2.7 shows the vertical and horizontal decision-making structures in a program hierarchy. In a vertical hierarchy in a program, the following are the main components:

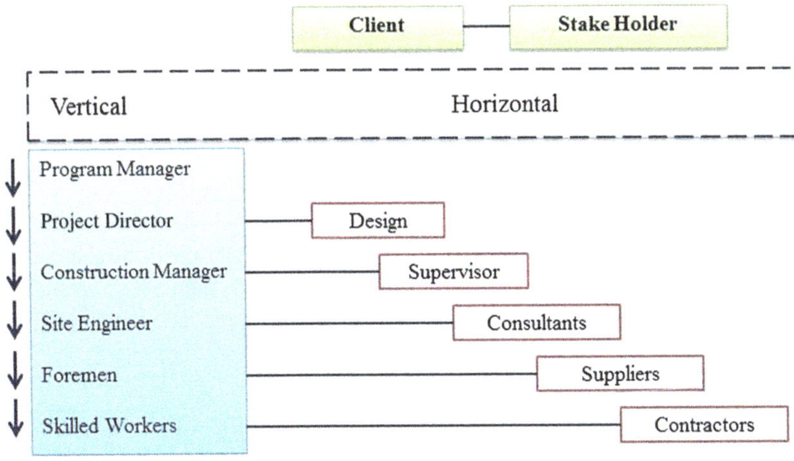


**Fig. 2.7** Interrelationship between vertical and horizontal scalar in program structure

1. Program manager, program director, or program leader;
2. Project managers;
3. Senior engineers;
4. Site engineers;
5. Technicians dealing with matters such as quality control and quality assurance, AutoCAD operators, quantity surveyors, surveyors, and safety officers;
6. Staff such as document controllers, security officers, attendance supervisors, secretaries, and office support staff;
7. Foremen;
8. Skilled laborers; and
9. Laborers.

In a horizontal hierarchy in a program, the following are the main components:

1. Client and stakeholders;
2. Contractors;
3. Designers;
4. Engineering consultants;
5. Other consultants in contracts, cost control, LEED, mechanical and electrical, value engineering, etc.;
6. Supervision team; and
7. Facility management.

## System Understanding—A Program Approach

A system is an organized, unitary environment composed of two or more interdependent parts, components, or subsystems and delineated by identifiable boundaries from its milieu. An engineering system consists of a large number of interconnected components, each of which may serve a different function, but all of which are intended for a common purpose. The degree of achieving the common goal is a measure of the system's effectiveness.

Every system is a sub-system of a yet larger system or component systems. No system is really independent of other systems, i.e., there are interactions between different systems. The state of a system at any moment is determined by the values of the relevant properties which the system has at that point in time. Any system has a large number of properties, only some of which are relevant to a particular purpose. The values of these properties constitute the state of the system.

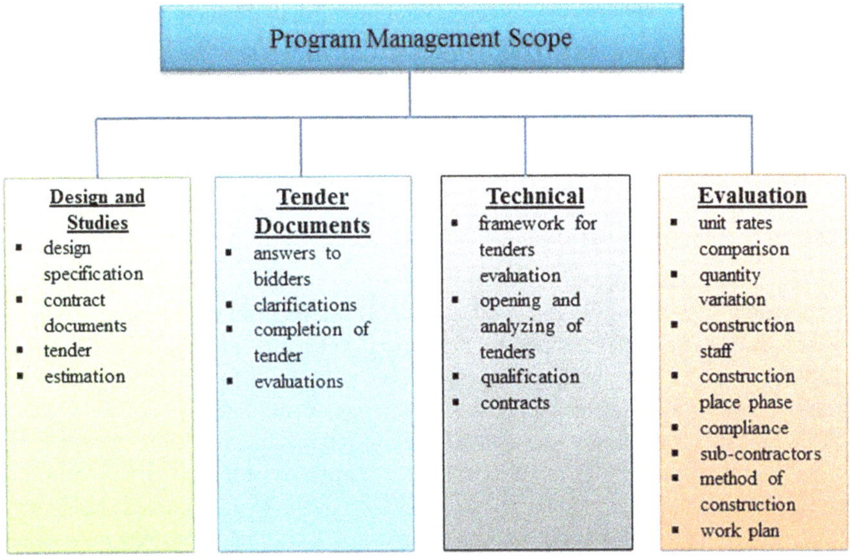


**Fig. 2.8** Typical program management system components

Basic components of a system for a typical program consist in principle of the following:

1. Engineering system, including design and constructability;
2. Type of system that relates to each project within the program;
3. Environment of a system;
4. Hierarchical system (A basic concept in systems thinking is that of hierarchical relationships between systems. A system is composed of subsystems of a lower order and is also part of a super-system. Thus, there is a hierarchy of the components of the system);
5. Systems analysis and decision making; and
6. Systems models including mathematical modeling, optimization techniques, statistical analysis, and intelligent models and simulations.

The structure of the systems analysis process for a program can be summarized in the following components as outlined in Fig. 2.8.

1. Project design and engineering;
2. Formulation of the planning and scheduling techniques;
3. Generation of alternative solutions for constructability;
4. Evaluation of alternatives;
5. Selection; and

6. Feedback. (The concept of feedback is important in understanding how a system maintains a steady state. Information concerning the outputs or the process of the system is fed back as an input into the system, perhaps leading to changes in the transformation process and/or future outputs.)

## System Environment

System environment comprises all other systems and their relevant properties which are not part of the system under consideration, but a change in any of them may affect the state of this system. The environment of a system includes also other systems that are affected by the system under consideration.

The system approach discourages the program manager from initially presenting a specific problem definition or adapting a particular solution to the problem; instead, the system approach emphasizes that the problem environment be defined in broad terms so that a wide variety of needs can be identified that have some relevance to the problem. These needs should reflect the complex relations and conflicts implicit in the problem environment.

## System Analysis

System analysis covers the comprehensive aspects of program management engineering practice and the application of modern decision analysis techniques in the planning and choice of engineering systems. The focus of system analysis is to optimize the use of resources (people, materials, money, and time). System analysis involves the application of many analytical tools such as utility and theory optimization, sensitivity analysis, accounting, knowledge base systems, and network techniques. Figure 2.9 shows a system analysis configuration.

The significance of systems analysis consists of the following:

1. Sharpening the program manager's awareness of the objectives of the program he or she is designing and planning. The program manager is required to make explicit statements of what the objectives are and their definitions;
2. Making precise forecasts;
3. Generating large alternatives;
4. Helping to make a decision; and
5. Suggesting strategies of decision making which can be used to select among possible alternatives.

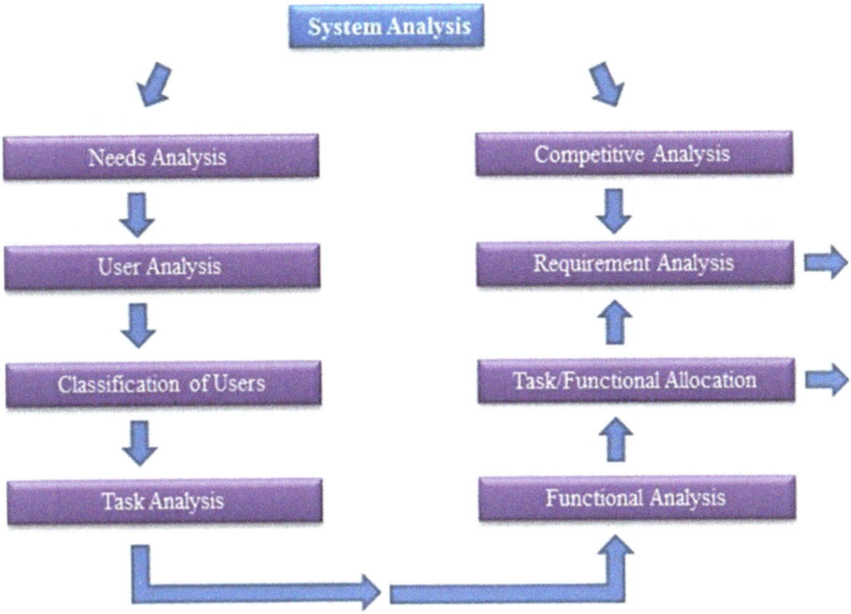


**Fig. 2.9** System analysis configuration

The fundamental steps in the structure of the systems analysis process are as follows:

1. Problem definition and statement of objectives;
2. Formulation of measures of effectiveness (MOE);
3. Generation of alternative solutions;
4. Evaluation of alternatives;
5. Selection and implementation; and
6. Feedback.

## System Models

These are abstract representations that describe the interactions between the complex factors of the program system environment and the causal dependencies among these factors so that the analysis can correctly perceive the effects of the substantial changes that may be introduced by large-scale projects. Refer to Fig. 2.10 for systems model building process.

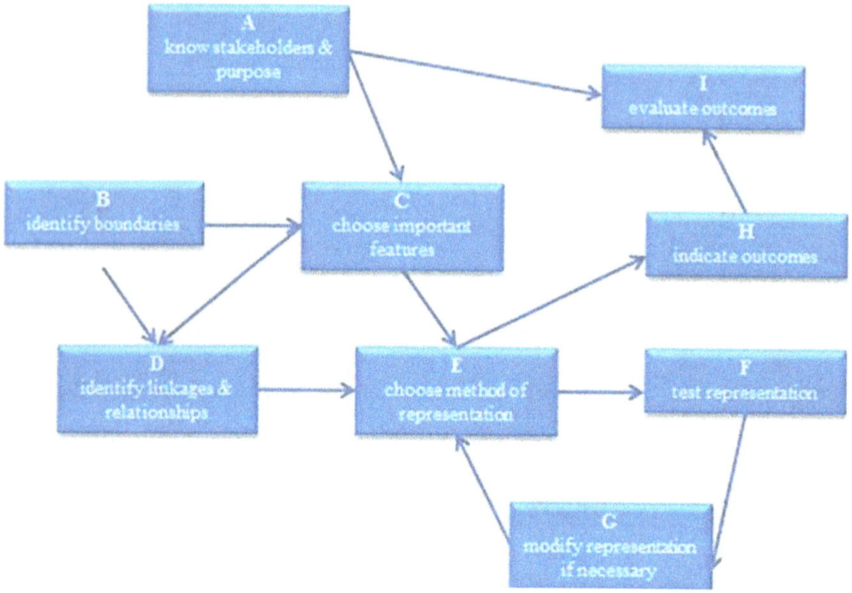


**Fig. 2.10** Systems model building process

The types of models vary as follows:

1. Iconic,
2. Analogue,
3. Mathematical or analytical,
4. Computer simulation, and
5. Artificial intelligence.

The systems model building process, achieving the above, includes as follows:

1. Model formulation,
2. Model verification (existing data),
3. Model application to predict new observations, and
4. Model refinement to achieve precision (Jackson 2000).

## Contingency View

The contingency view depends on a body of knowledge and research tasks that focus on interrelationships among key variables and projects in program management. It also emphasizes on the role of the program manager as diagnostician, pragmatist, and artist. The contingency view seeks to understand the interrelationships within and among projects as well as between the organization and its

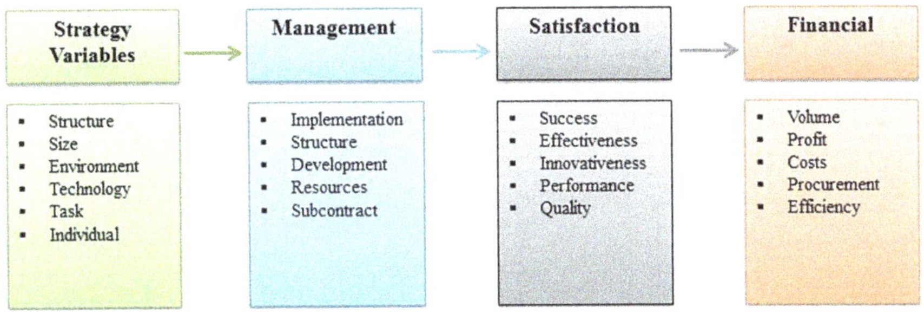


**Fig. 2.11** Contingency view components

environment and to define patterns of relationships or configurations of variables. It emphasizes the multivariate nature of projects and attempts to understand how program management operates under varying conditions and in specific circumstances. Contingency views and managerial actions are most appropriate for specific situations. Figure 2.11 shows the contingency view components which are divided into strategy, management, satisfaction, and financial.

This approach recognizes the complexity involved in managing new programs but uses the existing body of knowledge to relate the environment and the design, to match the structure and the technology, to integrate the strategy and the tactics, or to determine the appropriate degree of subordinate participation in the decision making, given a specific situation. Success in the art of program management depends on a reasonable success rate for actions taken in a probabilistic environment.

Contingency views represent a middle ground between the view that there are universal principles of organization and program management and the other view that each organization is unique and that each situation must be analyzed separately (Grandori 1984).

## Open and Closed Systems

Systems can be considered in two ways: (1) closed or (2) open. Open systems exchange information, energy, or material within their environments. Infrastructure and social development programs are inherently open systems. The closed system has rigid, impenetrable boundaries, whereas the open system has permeable boundaries between itself and a broader super-system. The boundaries set the domain of the organization activities. In a program comprising of residential buildings, the boundaries can be clearly identified. In an infrastructure program, the boundaries are not easily definable and are determined primarily by the functions and activities of the projects. Such an organization is characterized by rather vaguely formed, highly permeable boundaries. Figure 2.12 shows the advantages and disadvantages of open and closed systems.

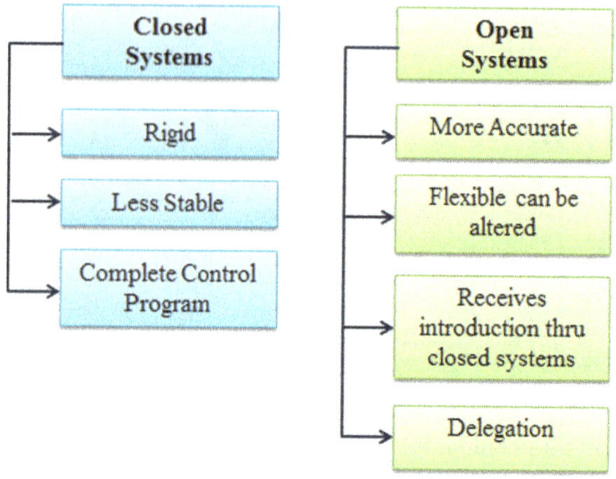


**Fig. 2.12** Open and closed systems—advantages and disadvantages

Many systems grow through internal elaboration. In the closed system, subject to design, planning, and constructability, the programs move toward entropy and disorganization. In contrast, open systems appear to have the opposite tendency and move in the direction of greater differentiation and a higher level of organization.

Traditional program management theories generally use a highly structured, closed-system approach. Modern theory has moved toward the open-system approach. The survival of the system would not be possible without continuous inflow, transformation, and outflow of information. The system must also receive sufficient input of resources to maintain its operation and also to export the transformed resources to the environment in sufficient quantity to continue the cycle.

For example, programs, including the construction of public buildings such as schools, hospitals, and colleges, receive inputs from society in the form of people, materials, money, and information and transform these into outputs of products, services, and structures. Finance and the market provide a mechanism for recycling of resources between the program management team and its environment. Also, even when we consider that the open system is the most suitable for program management topics, we should recognize that the concept of open or closed is a matter of degree. In an absolute sense, all systems are open or closed, depending on the point of reference. Thus, all systems are "closed" in some degree from external forces (McCullough 2008).

## Decision-Making Principles

Many principles are used to summarize the knowledge required for decision making in program management. They cover formulating a problem, obtaining information about it, selecting and applying methods, evaluating methods, and using decision-making techniques.

In this section, each principle is described along with its purpose, the conditions under which it is relevant, and the strength and sources of evidence. A checklist of principles is provided to assist in evaluating the decision-making process. The checklist can help one to find ways to improve this process and to avoid liability for poor selection, poor planning, and not providing the right information.

When program managers receive information, they often cannot judge its quality. Instead of focusing on the decision making, they decide whether the process is reasonable for the situation. Therefore, by examining decision-making processes and improving them, managers may increase accuracy and reduce costs.

It is crucial to separate the decision process from the analysis process. One possibility is to have one group do the planning and another do the analysis. Separating these functions could lead to reports showing different decisions for alternative plans. This principle is sensible and important, yet it is often ignored.

The program manager must describe how the decisions are to be made, and do so in intuitive terms. It may help to propose using a selection method on an experimental basis. The problem should be structured so that the program manager can use knowledge effectively for it to be useful for decision making. This will include identifying possible outcomes prior to making the decision. Determining possible outcomes is especially important for situations in which the outcomes are not obvious or in which a bias could lead to failure to consider a possible outcome. Brainstorming about possible outcomes assists in structuring the approach. For example, experts might be asked to brainstorm the possible outcomes from the imposition of an affirmative action plan in a workplace.

Other experts involved in a program, such as the designers, consultants, and specialists, should help to determine the prerequisite for a program specified by time, cost, specifications, constraints, and resources among other factors. Thus, program management can focus on the level of aggregation that yields the most accurate decision. As well as improving the use of program management by tailoring it to decisions, sufficient knowledge and information must exist to enable different levels of aggregation.

It is also essential to decompose the problem into parts. This will require the use of a bottom-up approach; that is, micro-managing each component, then combining them to improve the accuracy of decision making by improving reliability. Also, by decomposing the problem, a program manager can effectively use the alternative sources of information and the different methods. It is helpful to decompose the problem in situations involving high uncertainty and extreme (very large or very small) numbers.

The program manager must identify knowledge and information that might be useful in making a decision. While this should be guided by theory, the manager may need to be creative in seeking alternative types of knowledge and information.

It is also crucial to understand that information is critical as an input into the decision process. A positive correlation has been established between program performance and decision-making practice, and since a program is a temporary organization, a correlation between program performance and decision practices should be expected. The main challenges to a mega-project are inadequate, unreliable or misleading information, and the conflict between decision making, policy, and planning (Haidar et al. 2014).

## Improving the Accuracy of Decision Making

To follow this principle, program managers must have good prior knowledge of the problem to be dealt with. That knowledge can be based on the experience or research studies such as follows:

1. Received wisdom with little empirical testing. Received wisdom has been questioned, sometimes, in the belief that more information is always better;
2. Some researchers have ignored this principle in favor of knowledge and information mining, which assumes that the knowledge and information will reveal causal patterns;
3. Ensure that the information and knowledge match the situation;
4. Knowledge and information about past behavior in that situation are often the best predictors of future behavior;
5. Avoid biased knowledge and information sources; and
6. Avoid knowledge and information collected that are obviously biased to particular viewpoints.

Program managers must find alternative ways of measuring the same thing. If unbiased sources are not available, the manager may find sources with differing (and hopefully compensating) biases. For example, allocation of staff from project A to project B should equal the transfer of staff to project B from project A.

## Methodology and Knowledge Preparation

This is an essential part of the decision-making procedure and involves the program manager in preparing knowledge and information for the decision-making processes such as follows:

1. Clean up the knowledge and information;
2. Adjust for mistakes, changing definitions, missing values, and contingency. Keep a log to record adjustments; and
3. Use graphical displays for knowledge and information;
4. When judgment is involved, graphical displays may allow the program manager to better assess patterns, to identify mistakes, and to locate unusual events. However, experts might also be misled by graphs if they try to extend patterns from the past;
5. Program managers should be trained so that they do not try to match time patterns when making judgments in uncertain situations.

Program managers are required to select the most appropriate methods for making decisions. They can expect that more than one decision-making method will be useful for most problems. This will involve the following:

1. List all the important selection criteria before evaluating methods and
2. Accuracy is only one of many criteria. The relevant criteria should be specified at the start of the evaluation process.

Structured methods are those consisting of systematic and detailed steps that can be described and replicated. Structured methods are useful when accuracy is a key criterion and where the situation is complex.

Program managers are advised to select methods that are appropriate given the criteria, the availability, and type of knowledge and information. Prior knowledge, presence of conflict, and amount of change expected are also important. The selection of the most appropriate decision-making method, when alternative methods are feasible and there is much uncertainty, is summarized as follows:

1. Assess acceptability and understandability of methods to the consultants involved;
2. Ask project managers what information they need in order to accept a proposed method;
3. Examine the value of alternative methods; and
4. Examine whether the costs are low relative to potential benefits. Program managers seldom do this, primarily because of the difficulty of assessing benefits. This principle is unnecessary when potential savings are obviously large relative to the costs of the effective methods.

Program managers must try to keep decision-making methods simple as complex methods may include errors or mistakes that are difficult to detect. Simple methods are important when many managers participate in the planning and selection processes and when the stakeholders want to know how the decision was made. They are also important when uncertainty is high and little knowledge and information is available.

The decision-making methods should provide a realistic representation of the situation. Program managers should follow the following criteria:

1. Realize that they may have to add some complexity when developing optimization models;
2. Compare the matching of the method to the situation. This principle is most important when the match is not obvious. It is important when the situation is complex, as often happens for situations involving conflict among groups;
3. Be conservative in situations of high uncertainty or instability; and
4. Reduce changes to the extent that uncertainties and instabilities occur in the knowledge and information or in expectations about the future.

Some principles for decision making concern only judgmental methods. In general, program managers need to ask the right questions at the right time. These methods include as follows:

1. Pretest the questions you intend to use to elicit judgmental decisions;
2. Prior to collection of knowledge and information, questions should be tested on a sample of potential respondents to ensure that they are understood and that they relate to the objectives of the problem;
3. Frame questions in alternative ways;
4. The way the question is framed can affect the decision. Sometimes, even small changes in wording lead to substantial changes in responses;
5. Ask project managers to justify their decision making in writing; and
6. Support them in showing the reasons supporting their decisions.

Judgmental information can be combined with optimization methods and techniques in many ways to obtain the right decisions. This principle is important when the model used for decision making would not otherwise include judgmental knowledge. The use of this information as an input rather than to revise the decision is especially important when the decision could be subject to biases, as, for example, in scheduling and planning on the basis of the effects of new structural models where the program manager is more familiar with one system. The program manager, in order to combine a hybrid of empirical information, feed analysis, and optimization methods, must be consistent with the following:

1. Use structured procedures to integrate judgmental and quantitative methods;
2. Use prespecified rules to integrate judgment and quantitative approaches. In practice, analysts often violate this principle. The principle is relevant when you have useful information that is not incorporated in the optimization method. Whether to integrate will depend on the knowledge and information, types of method and expert information;
3. Use structured judgment as an input to optimization models;
4. Use judgment as an input to a model rather than revising the model's structure;
5. There is some empirical support and it challenges received wisdom;
6. Use pre-specified domain knowledge in selecting, identifying, and modifying the variables in the optimization methods; and
7. Subjective adjustments should be limited to situations in which you have domain knowledge that is independent of the model.

## Evaluation of Decision-Making Methods

When many solutions are needed for a particular situation, program managers should compare alternative methods of decision making. The comparison should include accuracy and other criteria. Among these other criteria, it is of particular importance to properly assess uncertainty. The principles for evaluating decision-making methods are based on generally accepted scientific procedures, namely:

1. Compare reasonable methods;
2. Use at least two methods, preferably including the current procedure as one of these. Exclude methods that would be considered unsuitable for the situation;
3. Whenever biases can affect the evaluation (which is often); knowledge of alternative approaches is helpful;
4. Use objective tests of assumptions;
5. Use quantitative approaches (statistics analysis, optimization techniques, knowledge-based systems, and genetic algorithms will be discussed in subsequent chapters) to test assumptions;
6. Design test situations to match the problem; and
7. Test decision-making methods by simulating their use in actual situations.

Presenting the outcome of the decision making is also crucial to improve the program manager's understanding and to reduce the likelihood of overconfidence. This process will include the following:

1. Present decision outcomes and supporting knowledge and information in a simple and understandable form;
2. Keep the presentation simple yet complete. For example, do not use insignificant digits because they imply false precision;
3. Graphs are often easier to understand than tables;
4. Clear presentations are especially important on the effects of program phase changes;
5. Provide complete, simple, and clear explanations of methods; and
6. Periodic assessments should be made to examine how the decisions are being used.

## Contextual Influences

Policy implementation, based on the decisions made, refers to the mechanisms, resources, and relationships that link the program execution policies to the program objectives. Understanding the nature of policy implementation is important because experience shows that policies, once adopted, are not always implemented as envisioned and do not necessarily achieve intended results. Moreover, some solutions and systems are provided with little attention as to how such activities fit into or contribute to broader program goals. Too often, policy assessments emphasize outputs (e.g., number of projects delivered) or outcomes (e.g., increased production in certain areas such as concrete activities) but neglect the policy implementation process which could shed light on barriers or facilitators of more effective implementation. Assessing the implementation process provides greater understanding of why programs work or do not work and the factors that contribute to program success.

Various factors influence the implementation of decisions, including their content, the nature of the implementation process, the parties involved in the process, and the context in which the policy is designed.

Program characteristics and contextual factors influence the program manager's approach to information feed and how challenges may be classified as "threats" or "opportunities". In particular, what a program manager perceives as important to senior management (an organizational context) is expected to influence his or her management priorities, hence decisions.

Literature on organizational behavior and decision making also infers that experience plays an important role in decisions and has a positive relationship with decision outcomes. So the program manager's professional experience (a personal context) could be expected to influence the information framework adopted on the program, hence the potential impact on the strategic outcomes. Figure 2.13 shows the different factors influencing a program manager decision implementation process.

The elements measured are the quantity, quality, and timeliness of information gathered by the program manager. These are combined to form the construct variable, information feed, derived from how the facts are constructed. These same subvariables could also be segregated as internally or externally focused information as a means of further sensing where problems may be coming from. The components of information feed include as follows:

1. Performance information on corporate financial services, HR management, and other performances;
2. Information on the "pulse" of internal and external stakeholders (stakeholder pulse factor);
3. Information on program efficiency, stakeholder management, benchmarks, etc. (project performance factor); and
4. Timeliness of information to the program manager toward decision making (information timeliness factor).

**Fig. 2.13** Factors that influence the implementation of decisions

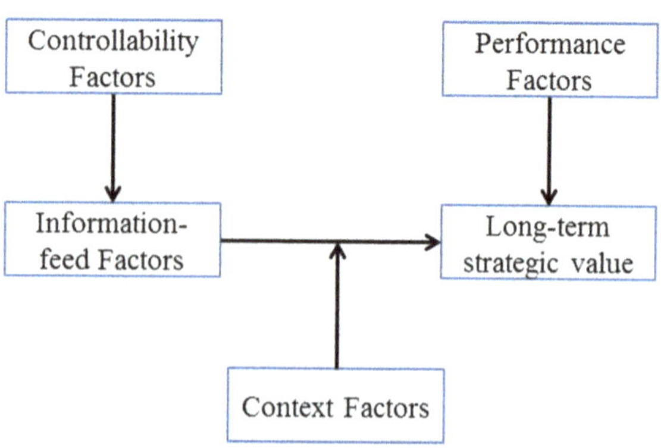

The contextual variable has two main components. First is the program manager's perception of what his or her senior management drivers are, for example, cost, schedule, stakeholder management, safety, quality, and economics. The second is information on the program manager's professional tenure, obtained as a measure of experience.

Strategic value is measured from the viewpoint of the program manager. Items measured are as follows:

1. Projects performance in comparison with objectives and aspirations of the host client and stakeholders on the program;
2. Health, safety, security, and environmental performance of the program;
3. Economic profitability; and
4. Making a significant socioeconomic contribution to society.

Integrating program performance information into a program manager's decisions has a positive influence on promoting the program value to the stakeholders, while exerting a negative influence of similar value to host communities who have their interest mainly tied to the benefits they expect to receive from the program (Gareth and Maynard 2013).

## Formalization Advantages

Program management formalization is directly connected to program success. Despite the merits of formalization, oversystematic and formalized systems may halt the progress of the program and increase organizational inertia as well as resistance to change. To understand the specific conditions that support the positive effects of formalization, it is essential to adopt a contingency perspective when investigating its effectiveness. Various characteristics, such as the size, complexity, or location of projects in the program, may influence the effectiveness of formalization. However, most studies do not take contingencies into account. Program complexity is of particular importance in the context of program management because larger programs and interdependencies between projects pose challenges for the manageability of programs. Figure 2.14 demonstrates the effect of single project management framework and its effect on formalization, program quality, and program success.

Formalization is defined here as the degree to which processes, procedures, work rules, and policies are clearly specified and followed. In program management, this includes the consistent use of defined procedures, methodologies, and tools. Formalization can take place at the level of single projects or at the program level. Established standards that have been developed explicitly for the program management domain describe processes and tools and provide guidelines and support to organizations in their application of management practices.

Formalization of processes helps to exploit economies of scale and of scope. Learning of processes becomes easier, coordination between processes is simpler,

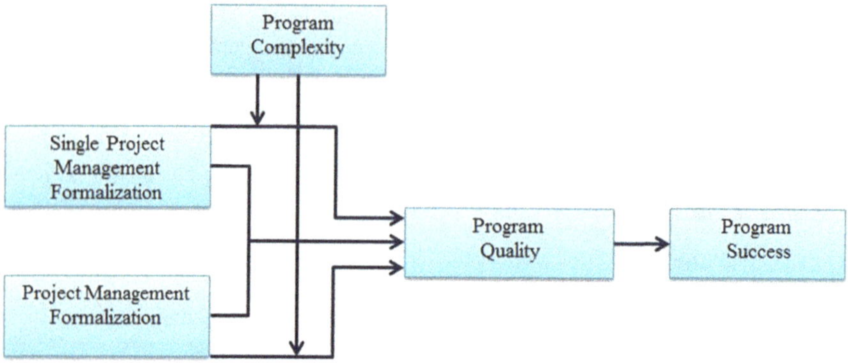


**Fig. 2.14** Relationship between formalization, program quality, and program success

processes become more reliable, and processes can be performed in a shorter time. Formal procedures include a shared and reproducible core process in which all project managers follow the same sequence of program phases, milestones, activities, and major deliverables for each program.

The benefit of consistently applied processes across a program is the ability to transfer process knowledge from one project to another. Through shared knowledge, program managers achieve a common understanding that is positively associated with information quality, thereby improving the speed and quality of communication within processes. A well-structured process provides predictability and control, and prevents malpractice by, for example, inhibiting the unjustified use of resources. Periodical program status reports and routine program reviews are beneficial for program tracking and initial program planning, which increases the percentage of projects completed on time.

Predictability of the scope, schedule, and cost of the program leads to higher transparency and reduces the residual performance risk, which increases performance. Furthermore, formalization can improve clarity in decision making. However, formalization is not always beneficial as, in specific scenarios with radical innovations, it may have negative effects; too much formalization may constrain creativity and interrupt innovative activities. While some authors argue that formalization has negative effects, in general, the positive effects of formalization prevail (Carpenter et al. 2013).

## Formalization Complexity

In contrast to single project management, program management is conducted at a higher hierarchical level. With an eye on the entire project program, a more holistic view is required to reflect previous experience, simultaneous projects, the organizational environment, and future organizational intention. Therefore, the exchange of information, management of resources, and coordination of the collection of projects become even more important for program. Various studies support the notion that the formalization of program processes significantly influences program performance.

In a decision tree (branch and node) procedure, after each process stage, a yes or no decision is made, and each project is assessed against pre-defined criteria to decide whether to continue with the same methods of execution for the project. If applicable, an action plan for the next stage is developed. At each branch, it is ensured that decisions and resource allocations reflect the needs of the entire program. These formal processes introduce structure, sequence, and clarity to all projects. Establishment of clear rules and guiding principles at the decision points lead to data integrity, and facilitate the comparison of divergent projects, ensuring that processes are comprehensive and responsibilities are well defined. Program process formalization, therefore, improves information and coordination quality by supporting interactions between different functional groups and projects, and facilitating interproject learning.

Formal procedures and rules enhance the availability and determine the format of information, thereby facilitating the comparison of diverse projects. For example, high levels of formalization in single projects result in clear resource requirements and a transparent planning and scheduling for these single projects. In turn, this increases the efficiency and speed of the formal resource allocation and prioritization process, and facilitates coordination between projects.

Without single project management formalization, the formalization of program management is elusive. Therefore, an increase in the formalization of program management without formal processes for single project management will be ineffective. However, formalization of single project management alone will not be effective either because it lacks a holistic view. Furthermore, the definition and implementation of a formalized program process will increase, and reinforce the formalization of the single project process. While single project management improves efficiency, program management enables organizations to increase their effectiveness. Simultaneous single project management formalization and project program management formalization increase the positive effect on program quality (Byrne 1998).

## Program Complexity as a Contingency Factor

The optimal degree of formalization depends on the characteristics of the task, which are a core theme in task-related contingency theories; hence, it is necessary to adopt a contingency perspective to specify the conditions under which formalization becomes more or less desirable and effective. Two kinds of task related contingency theory have been developed in relation to the impact of formalization. The first uses the complexity of a task as a moderating factor, and the second uses the uncertainty, risk, or innovativeness of a task.

In program management, the systemic perspective is often used to describe the complexity of a program which is defined as the size of the project program and the degree of interdependency between projects. This perspective includes the following determinants of complexity:

1. The number of projects;
2. The degree of interdependency between the projects; and
3. The magnitude and predictability of changes in the projects and interdependencies.

Similar arguments have been used to define the complexity of single projects. Because the magnitude and predictability of changes in the projects are also central elements of uncertainty, the size of a program, and the degree of interdependency between the projects are measures of complexity. The more these interdependencies occur, the higher is the complexity of a program. Projects in a program may be linked by outcome, resource, or knowledge interdependencies as such:

1. Outcome interdependency occurs when one project uses the resources of another project,
2. Resource interdependency occurs when different projects concurrently compete for the same resources, and
3. Knowledge interdependency occurs when the knowledge generated in one project is relevant for another project.

Any collection of interrelated projects requires coordination of project management activities. The need for coordination results from the inevitable effect of changes in one individual project on the execution of another project in the program. For example, delays in one project place the resources availability of the entire program at risk when projects share the same scarce resources. Therefore, with increasing program size and stronger project interdependency, coordination becomes even more important. Because formalization enables better coordination, it may be especially beneficial in programs with high complexity. Program complexity also increases the opportunity to leverage synergies into knowledge, technological platforms, and end users. Resource conflicts become more likely and the allocation of resources becomes more challenging.

## Decision-Making Formalization Methods

Typically, the program manager is responsible for the immediate management of the program as well as conceptual and advisory activities to shape the program processes. Thus, the program manager is in a unique position to judge the applied procedures, methods, and processes for managing the program. Although program managers can be considered the best source for the variables, the chosen key is that there are no right or wrong answers. The key methods for formalization include the following:

1. The program is consistently aligned with the firm's future;
2. Firm strategy is implemented by the program in an optimal way;
3. The program resource allocation reflects strategic objectives;
4. The program has a good balance between opportunities and risks;
5. Transparency is important;
6. Accessibility to all relevant information on a project's status is made easily and quickly;
7. Presentation of information on the program is standardized at the top management level;
8. Program managers are continuously provided with relevant information on the entire program;
9. Program status and resource information can be interpreted easily and quickly;
10. Resource information is delivered as is necessary for decision making;
11. A detailed plan is provided for each project;
12. Each project gets assigned a defined project budget within the program;
13. Program monitoring takes place continuously for the whole duration of a program;
14. Program progress is regularly tracked, as well as completely and routinely recorded, for each project within the program;
15. Program management process is divided into several phases;
16. All process phases are concluded by an explicit approval gate;
17. Program management process is precisely specified;
18. During a program review, all projects are rigorously examined;
19. A shared understanding of the program management process is reflected in the activities of all projects;
20. A very structured program management process is implemented;
21. A high degree of alignment between projects is required with respect to the scope of each;
22. The output of one project is often part of another project or a component of the whole program;
23. Scope changes of individual projects impact on the execution of other projects; and
24. Often projects can only be continued if the precise results of other projects are known.

## Reliability and Validity of Decision Making

Reliability is the consistency of measurement, or the degree to which an instrument measures the same way each time it is used under the same conditions with the same subjects. In short, it is the repeatability of your measurement. Reliability is usually estimated as test/retest and internal consistency.

Test/retest is the more conservative method to estimate reliability. Simply put, the idea behind test/retest is that you should get the same score on test 1 as you do on test 2. The three main components of this method are as follows:

1. use your measurement instrument at two separate times for each subject,
2. compute the correlation between the two separate measurements, and
3. assume there is no change in the underlying condition (or trait you are trying to measure) between test 1 and test 2.

Internal consistency estimates reliability by grouping questions in a questionnaire that measures the same concept. For example, one could write two sets of three questions that measure the same concept and, after collecting the responses, run a correlation between those two groups of three questions to determine whether the instrument is reliably measuring that concept.

The primary difference between test/retest and internal consistency estimates of reliability is that test/retest involves two uses of the measurement instrument, whereas the internal consistency method involves only one use of that instrument.

Validity is the strength of our conclusions, inferences, or propositions. More formally, it can be described as the best available approximation to the truth or falsity of a given inference, proposition, or conclusion. There are two types of validity commonly examined in program management:

1. Internal validity asks if there is a relationship between the program plan and the outcome; in other words, if it is a causal relationship or not; and
2. External validity refers to our ability to generalize the results of our study to other settings.

The real difference between reliability and validity is mostly a matter of definition. Reliability estimates the consistency of the measurement, or more simply the degree to which an instrument measures the same way each time it is used under the same conditions and with the same subjects. Validity, on the other hand, involves the degree to which we are measuring what we are supposed to, or, more simply, the accuracy of the measurement. Many scholars believe that validity is more important than reliability because if an instrument does not accurately measure what it is supposed to, there is no reason to use it even if it measures consistently (reliably).

## References

Byrne, David. (1998). *Complexity theory and the social sciences*. London: Routledgae.

Carpenter, M., Bauer, T., Erdogan, B., Short, J. (2013). *Principles of management*, version 1.1, Nyack, NY: Flat World Knowledge.

Dillon, S. M. (1998). Descriptive decision making: Comparing theory with practice New Zealand. *Department of Management Systems, University of Waikato New Zealand*. http://orsnz.org.nz/conf33/papers/p61.pdf.

Eweje, J., Turner, R., Müller, R. (2012). Maximizing strategic value from megaprojects: The influence of information-feed on decision-making by the project manager. *International Journal of Project Management*. (Impact Factor: 1.53). 08/2012.

Grandori, A. (June 1984) A prescriptive contingency view of organizational decision making. *Administrative Science Quarterly*, *29*(2), 192–209. Published by Sage Publications, Inc.

Haidar, Ali D., Wells, Kenneth, & Thomas, Peter. (December 7, 2014). *Programme management in construction hardcover—December 7, 2014*. London: ICE Publishing.

Jackson, M. (November 30, 2000). *Systems approaches to management Paperback—November 30, 2000*, Biddles. UK: IBT Global UK Ltd.

McCullough, C. (2008). *Open and closed systems analysis*. http://www.academia.edu/5698820/Open_and_Closed_System_analysis.

Naidu, N. V. R., Krihsna Rao, T (August 21, 2008). *Management and entrepreneurship—Paperback—August 21, 2008*. I.K. International Publishing House Pvt. Ltd.

Shepherd, G. N., Rudd, J. M. (2013). The influence of context on the strategic decision-making process: A review of the literature. *International Journal of Management Reviews*, *16*(3), 340–364.

Wakker, P. P. (August 23, 2010). *Prospect theory: for risk and ambiguity Paperback—August 23, 2010*. Cambridge University Press: Cambridge.

# Chapter 3
# Planning and Scheduling—A Practical and Legal Approach

**Abstract** Construction, a complex process that can encounter many disruptions and unexpected conditions, needs extensive tools that are capable of handling disputes and informing the parties of their remedial obligations. Network-based scheduling techniques and programming of the complex sequences of activities and their dependences offer an invaluable tool for assessing delays, negotiating timely settlement of variations and solving disputes throughout the life of a construction program. Since time is a critical element in the construction process, owners and contractors risk incurring additional and substantial costs when construction programs are finished beyond their contractual completion dates. This chapter will look at the principles of scheduling and planning and the applications of the Critical Path Method in regards to liability, the legal framework, delays, extension of times, and Standard Forms of Contract. A well-structured Critical Path Method serves as a road map, plotting out a logical succession of steps, from which a series of smaller, more specific tasks emanates to deliver the projects comprising the program. It also serves as a script showing the interactions of key participants in the projects. This chapter will look at the fundamentals of planning and scheduling methods, and the importance of the application of the Critical Path Method as a well-prepared program schedule can make the difference between a program that progresses smoothly and one that is characterized by delays and other disruptions.

## Introduction

One of the program manager's principal roles is to monitor the Critical Path Method. He must understand how the laws and arbitration courts view the Critical Path Method in case of delays. Also, in this chapter, we will look closely at how the Standard Forms of Contract approach the Critical Path Method and include it in their clauses, since the traditional forms of contract make little contractual provision to integrate the programming of activities into structural obligations for construction projects. In many projects, and indeed programs, delay and disruption issues

 
A.D. Haidar, *Construction Program Management – Decision Making and Optimization Techniques*, DOI 10.1007/978-3-319-20774-2_3

that ought to be managed within the program all too often become disputes that have to be decided only after the delays have occurred and disputes have arisen (Gerrard 2007).

A properly working schedule is able to address many of the problems with respect to time, delay, and causation; yet there is often reluctance by the employer, or by the consultants that advise him, to allow the initial tender program the status of a contract document. This reluctance often stems from the fear that the contractor is more proficient in the use of programming techniques and, therefore, is able to use the program to his advantage and conversely to the employer's disadvantage. As a result, the opportunity to make use of the program to analyze post-contract time-related disputes is lost and disputes are more likely to ensue.

The resolution of disputes on large construction and engineering contracts increasingly involves the use of computer-based delay analysis techniques to assist in the identification of the cause of critical delay to a project or a program and, in the more sophisticated cases, to assist in the computation of claims for lost productivity. While the industry is becoming more and more familiar with the use of the tools and techniques employed in the process of delay analysis, the future will bring more common agreement upon their correct application (Schumacher 1996).

The problems of unresolved delays and disruptions in construction contracts are notorious. Unintended delays cause disputes and losses in construction and civil engineering contracts worldwide, and the common view in the industry is that many disputes arise because the parties do not understand the way delays occur and how their consequences could be avoided.

Programming of the complex sequences of activities and their dependence is one of the principal skills of the successful program manager. All but the most simple of projects will proceed from such a program. If allegations of delay are to be shown effectively by the contractor and considered properly by the contract manager administering the program contract, it will be found that, in most construction programs, a properly prepared Critical Path Method indicating quantity output, design stages, procurement structure and progress, resources allocations, physical progress of each project within the program, milestone completion dates, and the passage of time is essential.

The Critical Path Method, when married to law, must persuasively demonstrate the desired and sought for result by virtue of the justice, equity, and fairness of each party's position.

As the courts and the arbitrators become more familiar with delay analysis techniques, it is likely that there will be an increasing number of reported cases addressing these issues, giving guidance to delay analysts as to the preferred approaches to take and censuring experts who fail to present cogent and balanced evidence that assists the court in its decision making.

## Scheduling and Planning Development

Schedules must be reasonably accurate and contain a level of detail appropriate for each project life cycle within the program. They must be easy to read and understood by all program participants. And they must reflect the owner's and stakeholders' requirements by identifying program milestone dates, including the feasibility study, master plan, estimated time of preliminary studies, site acquisition, design, the selected commissioning process, consultant appointment including the supervision team, bidding, negotiating, procurement, construction, occupancy, and facility management.

A good schedule clearly identifies the program delivery requirements and provides a tool for managing each project within the program. Architects and engineering consultants, design builders, contractors, and the program director can work with the owner in each stage of the program to refine the program schedule. Figure 3.1 shows the key components of a program schedule.

The program schedule developed during the program conception differs from the construction schedule. It is, first of all, substantially broader. The program schedule is first formulated during this conception stage and often contains tasks, such as feasibility study, financial approach including cash flow study, acquisitions, joint ventures, consortiums and partnerships, planning and scheduling methodologies, appointment of the consultant teams, site studies and site availability, design, and other activities that must be completed before the physical construction is begun. The schedule may then progress many years beyond the program completion according to the program needs. The construction schedule is just one part of the overall program schedule, albeit an important part. Precise dates for project milestones may not be possible to determine during conception of the program (Wickwire et al. 1999).

As program conception tasks are completed, dates may become clearer to the key participants and more detail may be added which will include milestones, which are important dates or accomplishments within the program time plan. The schedule is also separated into tasks, along with the duration of these tasks. The tasks may be divided into subtasks or the series of steps required before a task may be completed. Figure 3.2 shows the program caption schedule and the main activities that affect its proper realization.

**Fig. 3.1** Key components of a program schedule

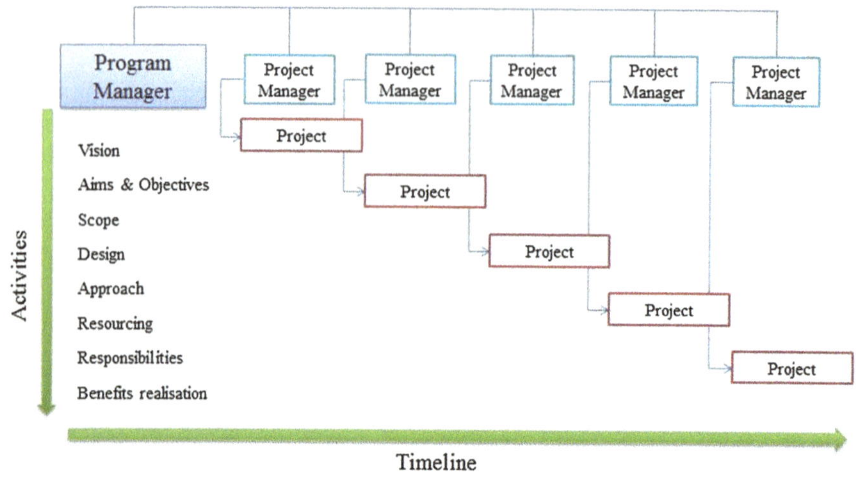


**Fig. 3.2** Program conception schedule

The concept schedule will be developed into a detailed and more comprehensive schedule, often called the development schedule or the overall program schedule. This schedule will include the timely preparation of contract documents, labor availability and training, procurement, and types and allocation of resources. The overall program schedule must identify decisions to be made and milestones to be achieved at each stage of the program and the ramifications when the milestones are not met.

The overall program schedule must identify the decisions to be reached and key deliverables at each critical project stage and the potential effects when these milestones are not met. Deliverables are tasked items or components, performed or provided, to complete each project design stage of the program. Deliverables allow milestones to be achieved with timely decisions. They should identify decisions, actions, and approvals that must be made and the time periods allotted for them.

The program manager can assist the client in producing a realistic program schedule, using experience and professional tools such as scheduling software that can add value and efficiency to the process. As with all other studies and tasks performed during the program conception, the services of the consultant bring value to the program as the client will not have the right expertise to deal with the wide scope and many facets of a program, even if repetitive. The program manager's role is always critical for the successful construction of a program.

Depending on a program delivery method, milestones of an overall program schedule developed include durations for:

- Feasibility and finance
- Predesign activities
- Design activities
- Consultant appointment

- Site acquisition
- Program delivery and team selection
- Processes, including requests for qualifications, interviews, and negotiation of agreements
- Design and construction documents and deliverables required at completion of each phase and stage
- Approval processes such as plan review and permit, zoning approval, and approvals of other regulatory agencies
- Bidding, negotiating, and construction contract award
- Construction stage activities, including lead times for major activities, components, systems, and subsystems
- Program commissioning activities
- Contract completion and occupancy activities.

The following factors (among others) require the inclusion of scheduling contingencies to allow adjustments to the schedule without totally disrupting the concept or overall project duration:

- Discovery of new program information, including unexpected results of site investigations
- Changes in laws, regulations, and codes that affect design or construction
- Length of approval processes
- Social factors, such as labor strikes or political events
- Economic factors, such as the state of the economy, the prevailing lending rates, and availability of key products and components
- The effects of geographical locations.

Schedules can be developed by a variety of approaches. One approach provides for the establishment of a required date of delivery or use for the program and then

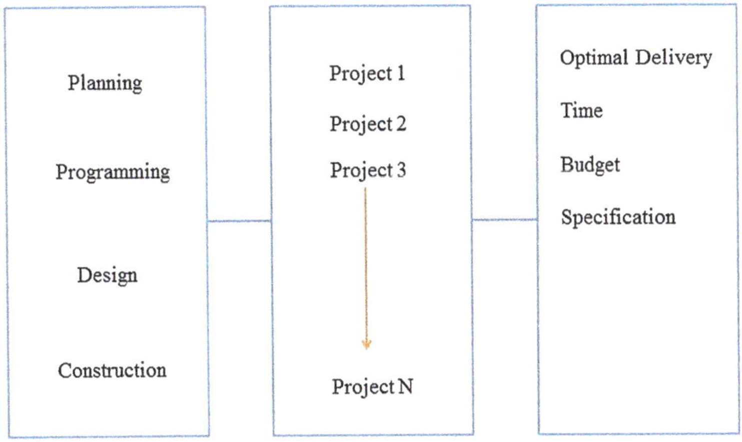


**Fig. 3.3** Scheduling and planning main components and interrelationships

identifying all preceding milestone dates by working backward from this date. This approach is utilized when the completion date is the most important scheduling criterion. Figure 3.3 shows the interrelationships between the main scheduling and planning components.

## Critical Path Method Overview

Program management for the construction industry, a complex process that can encounter many disruptions and unexpected conditions, needs extensive rules and planning methods which must inform the parties of their remedial obligations, handle disputes, and provide clear programming of the complex sequences of activities and their dependence.

The critical path method is a tool that demonstrates the shortest possible path to completion at any stage by breaking down the interrelationship of the discrete elements that comprise the activities to be undertaken. A critical path method determines activities that are dependent upon each other and identifies the longest path for the completion of those activities. In this approach, the schedule is dictated by the activities that are on the critical path of the completion date of the program. As the program progresses, those managing the schedule can compare actual versus predicted performance times for each step and adjust the rest of the schedule accordingly. Figure 3.4 shows critical path method structure for a program.

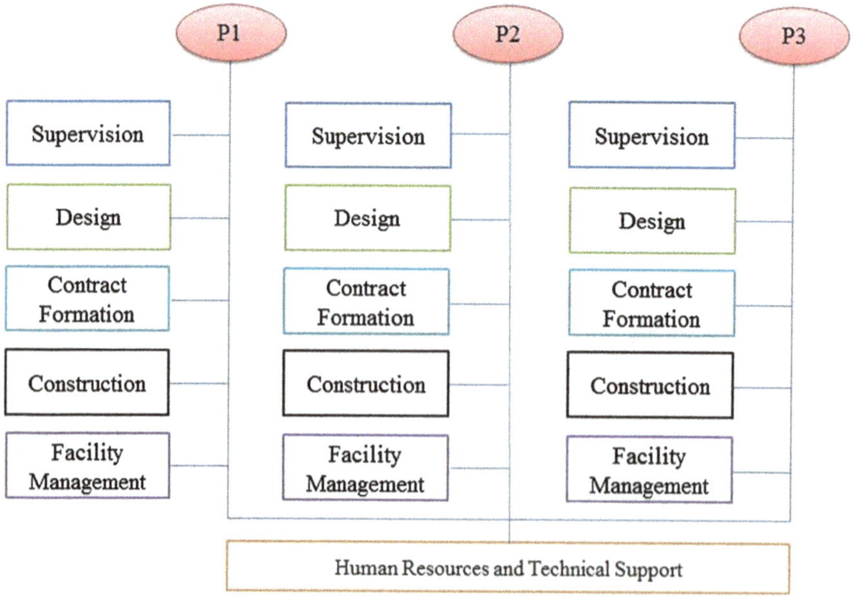


**Fig. 3.4** Critical path method structure for a program

Therefore, the critical path method provides a tool by which actual job progress against a plan is monitored, thus enabling an early alert of actual and potential delays which could adversely affect the program completion date. A well-constructed program using the critical path method allows the parties to identify which activities, i.e., parts of the program, are critical. Any delays to the activities falling on the critical path are likely to cause delay to the completion date.

In summary, a critical path analysis is regarded as a model that approximates the sequence and duration of the operations and activities for a program. The perfect model would follow each resource for the program and show what it was doing and in what location, and the sequence in which it carried out its work. The main activities to be considered for creating a critical path for a program are as follows:

1. Financing and cash availability
2. Design
    - Appointment of architects and design team
    - Master plan
    - Prototype design
    - Concept design
    - Schematic design
    - Design development.
3. Contract formation and preconstruction studies
    - Appointment of consultants and quantity surveying team
    - Composing standard contracts for infrastructure and common works
    - Contract composition for the different projects/packages
    - Finalizing bill of quantity.
4. Selection of contractor(s)
    - Prequalification of contractors
    - Selection of contractors
    - Technical and cost submittals
    - Bid bonds
    - Analyzing tenders and selection of contractors.
5. Construction phase of the projects comprising the program
    - Performance bond and bank guarantee submittal by contractors
    - Mobilization of contractors
    - Construction phase
    - Testing and commissioning
    - Completion and handover.
6. Human resources
    - Recruiting of engineers and staff
    - Recruiting of labor

- Equipment distribution
- Insurance and administration of projects.

7. Technical support

   - Planning and scheduling
   - Shop drawings
   - Quantity surveying
   - Invoicing
   - Quality assurance
   - Quality control
   - Value engineering
   - Health, safety, and environment.

8. Supervision

   - Appointment of engineer
   - Mobilization of staff to different sites.

9. Facility management.

## Activity Duration

An activity duration is the amount of time estimated for its completion. The time units for the program are usually calendar days, weeks, or months. Fractional time units can be used for activity durations, but integer time units are more common. The only requirement is that the use of the time units expressed should be consistent throughout the schedule.

The activity duration is a function of the estimated quantity of work that must be accomplished and the average production rate at which that work can be accomplished. Basically, the activity duration can be estimated by the following Eq. 3.1:

$$\text{Activity duration} = \frac{\text{Work quantity}}{\text{Production rate}} \tag{3.1}$$

Program activity durations are estimated, and usually, it is not essential for these estimates to be consistently exact. If all durations are reasonable, then variations in activity durations will compensate one another, resulting in reasonably accurate program duration. The accuracy in estimating activity time should not be overemphasized because doing so could make the task of developing reliable duration estimates unnecessarily complicated.

To obtain the best estimate of each activity duration, the scheduler must consult with the program manager. The accuracy of duration estimates depends on many factors including:

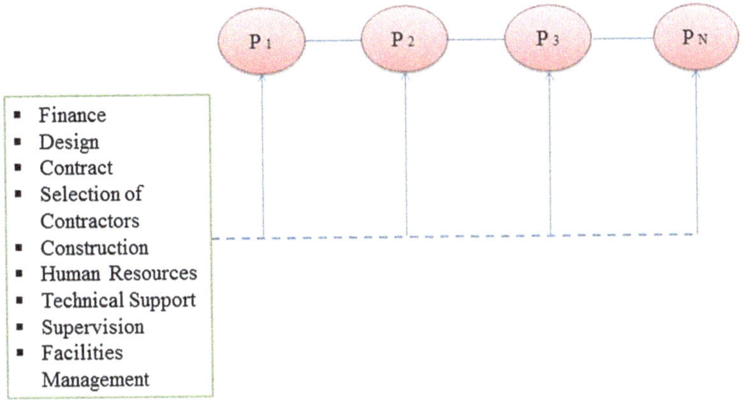


**Fig. 3.5** An example of activity interactions

1. Number of projects
2. Nature or complexity of the program
3. Design duration
4. Methods of construction
5. Resource availability
6. Work quantity
7. Labor and equipment productivity
8. Quality of field management
9. Geographical locations and site conditions
10. Concurrent activities (Thomas 2004).

To gain the best duration estimate, it is also necessary to obtain duration estimates from each project manager involved in the program and then to incorporate these estimates into the schedule. Figure 3.5 shows an example of activity interactions.

## Logical Relationship

A major component of schedule preparation is to determine the sequence or develop the activities logic. After estimating durations, the next step in preparing the schedule is to arrange activities in the order in which they should be performed, and in other words, to define the logic as the order in which the activities will be performed and logical relationship as the relationship of any activity to another. Figure 3.6 shows the types of logical relationships and their difference between a project and a program.

There are three possible logical relationships among activities, namely:

1. Predecessor
2. Successor
3. Concurrent or independent.

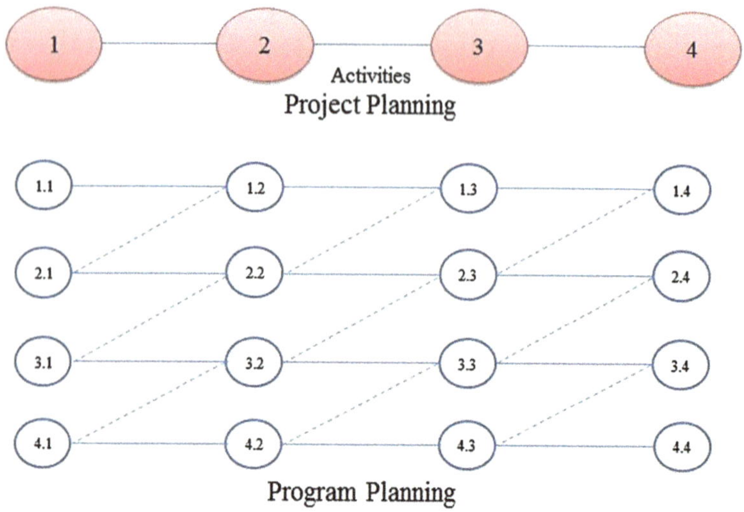


**Fig. 3.6** Types of logical relationship

A predecessor relationship between two activities means that one activity must be completed before the other activity can begin. For instance, mobilization must occur before the method statement for mobilization is made, and building work cannot commence unless design development is completed. Conversely, a successor relationship between two activities means one must come after the other. If two activities have neither a predecessor nor a successor relationship, then both activities are independent of, or concurrent with, each other. A concurrent or independent relationship between two activities does not always mean that the activities will be performed at the same time. It simply means that the completion of one activity is not contingent upon the beginning or completion of another.

Activities logic must be logical, reasonable, and possible to carry out through the program construction. If the sequence is unreasonable, irrational, or impossible, the schedule is considered to be unworkable.

Two important aspects for determining sequence on the logic diagram should be highlighted. The first aspect is that proper sequencing requires an understanding of the program construction itself, not merely its scheduling. The second aspect is that there is always more than one correct way to sequence any construction program. The task of sequencing a schedule requires experience and knowledge of the program management issues so that activities can be arranged in the sequence most likely to result in cost-efficient and timely completion of the program (Antill and Woodhead 1990).

## Forward Pass and Backward Pass

The next step after estimating activity durations and establishing the logic is to calculate the critical or longest path. This calculation involves determining four event times for each activity: Early Start (ES), Late Start (LS), Early Finish (EF), and Late Finish (LF). Figure 3.7 shows an example of a CPM showing ES, LS, EF, and LF.

Early Start and Early Finish times for each activity can be determined through the process known as the Forward Pass. In performing the Forward Pass calculation, all activities in the network are assumed to start as early as possible. Early time calculations begin at the first event—which has an Early Start time of zero—and work forward from there. The value of all subsequent early event times is the sum of the event time and the activity duration, obtained through the following equation:

$$\mathrm{EF} = \mathrm{ES} + D \tag{3.2}$$

where
$D$ = Estimated original durations

When two or more activities merge into a node, the value of the early time is calculated for each path using the above Eq. 3.2, and the largest value is used in Eq. 3.3:

$$\mathrm{ES} = \text{Maximum EF of direct preceding activity(s).} \tag{3.3}$$

The Backward Pass process calculates the latest time that each event in the network can start and finish without delaying scheduled completion, as figured by the Forward Pass. In performing the Backward Pass calculation, all activities are assumed to start and finish as late as possible. The calculation is similar to the Forward Pass calculation but is calculated from the end to the beginning of the project; it starts with the last activity in the network and works backward.

The Backward Pass calculation begins with the Late Finish of the last activity in the network, equal to either an arbitrary scheduled completion time or the Early Finish time calculated from the Forward Pass. The late time of succeeding events is determined by deducting the activity's duration from the late event time. Therefore, the Late Start of an activity can be calculated as follows in Eqs. 3.4 and 3.5:

$$\mathrm{LS} = \mathrm{LF} - D \tag{3.4}$$

The smallest of all calculated values for a particular activity is the late event time when the activity involves merging two or more activities:

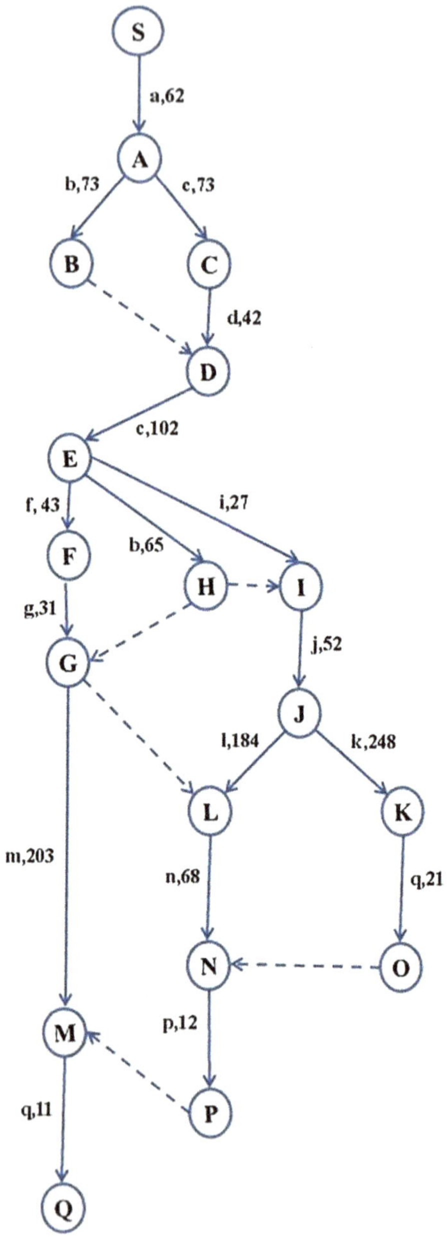


**Fig. 3.7** An example of a CPM showing ES, LS, EF, and LF

$$\text{LF} = \text{Minimum LS of immediate subsequent activities.} \tag{3.5}$$

In summary, assigning times for completing each project activity enables the linear graph to serve as a scheduling model. Thus, it is possible to determine the following:

1. The earliest time at which an activity may be started (known as earliest start time or ES). ES represents the earliest time that a given activity can begin after the initiation of a program. This time is a function of the other activities that must be completed prior to starting the particular activity under consideration. The earliest start time can be ascertained by determining the maximum necessary time required for preceding activities. This requires summing the time requirements along each linear graph path from the starting point to the activity involved.
2. The minimum time in which the total program may be completed. This is given by the duration of the maximum earliest start time path. That earliest start time path from the start of the program which determines the project completion time is known as the critical path.
3. The latest time at which an activity may be started if the program is to be completed in a minimum time (known as latest start time or LS). LS is specified in terms of the time from the start of the program; however, it is computed backward in time from the program finish node based on the minimum program time.
4. The earliest time at which an activity may be finished (known as the earliest finish time or EF).
5. The latest time at which an activity may be finished if the program is to be completed in a minimum time (known as the latest finish time or LF) (Antill and Woodhead 1990).

## Critical Path

After the Forward Pass and Backward Pass calculations have been made, some activities will be found to possess the same early and late times. These activities must begin and end on their early start and finish times, as failure to complete them within these limits can affect the completion time of the entire program. Thus, these activities are called critical activities.

These critical activities form a continuous chain through the network known as the critical path: the longest path from the beginning to the end of the program. Various non-critical activities not on this path have a certain amount of float time, and any delay in them within this predetermined time will not delay project completion. As a result, program managers, logically, focus most of their attention to critical activities. Figure 3.8 shows an example of critical path activities.

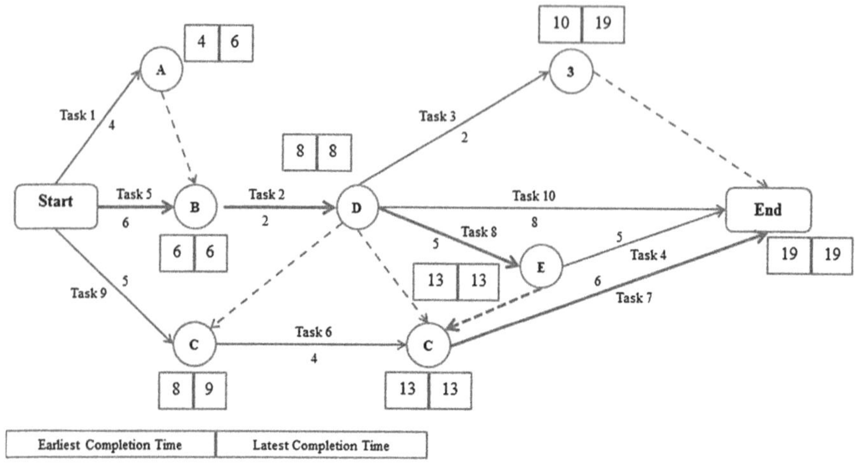


**Fig. 3.8** Example of critical path activities

## Advantages and Disadvantages of the Critical Path Method

The Critical Path Method possesses numerous advantages, as well as disadvantages.

**Advantages**

- **Minimizing erroneous and misleading schedules**. Because preparation for scheduling the Critical Path Method requires detailed analysis, the scheduler usually works alongside the program manager, who has a better understanding of the program. This requirement minimizes the possibility of erroneous or misleading schedules
- **Well established and easy to understand**. The Critical Path Method is well established and easy to understand, with techniques for drawing and calculating program durations developed from advanced high technologies. The Critical Path Method technique is like a common language, widely used in and accepted by the construction industry
- **Accepted in proving delays**. The Critical Path Method can be used to determine the length of delays or additional times needed as a result of unexpected events occurring or changes demanded during the construction process of the program. Review boards and courts of law have accepted the use of the Critical Path Method technique to prove delays and inefficient claims, identify the causes of such delays and inefficiencies, and assign responsibility for them. However, when used in delay analysis, the schedules should be realistic and reasonable.

**Disadvantages**

- **Simple nature of Critical Path Method logical relationship**. The simplicity of the Critical Path Method's logical relationship makes it difficult to specify in reality such relationships. The single logical relationship used in the Critical Path Method technique simply is not adequate for expressing all of the various sophisticated relationships possible in the program. In the case that activity B can start when activity A is half finished, for instance, the single logical relationship in the Critical Path Method cannot specify the real relationship in a simple way; two subdivided activities are needed to replace activity A to present the correct relationship between the activities A and B. In summary, in order to replicate the program construction activities, this method requires many subdivided activities linked as "stair steps," where activities A and B can start or finish together or the start of activity B may overlap the completion of activity A

- **Reliability of activity durations**. The reliability of activity times computed from the Forward Pass and Backward Pass calculations is questionable. The activity times are based on the network logic or sequence and on estimated activity durations developed by the planners. In most cases, the estimated task durations will not prove to be exactly correct, nor will the tasks be performed in the exact sequence. Hence, it is uncertain that the actual activity times will correspond exactly with the scheduled activity times. It is more likely that the Critical Path Method durations are overrun rather than underrun, although contingencies are considered and incorporated during the scheduling preparation
- **Tardiness of the Critical Path Method scheduling**. The problem is a simple one. It requires time for managers and decision makers to review status reports. By the time they review them, the information they obtain tends to be outdated. In addition, the practical integration of the Critical Path Method based progress and cost control is extremely difficult, non-productive, and expensive
- **Attention of program managers to critical activities**. Construction planning involves not just paying attention to a particular path related only to activity durations; it also involves giving equal attention to all activities in the network and perhaps alerting program managers to delete such artificial paths. Cost is also a significant element in construction planning; therefore, it should be considered in scheduling and in determining the criticality of activities (Kallo 1996).

## Float

As a management tool in construction projects, the Critical Path Method scheduling technique is well known and widely accepted. In fact, many construction contracts, especially those written for large public and private programs, require contractors to submit and routinely update the Critical Path Method to show critical and non-critical activities. In such programs, another occurrence is also common place in that many participating parties attempt to appropriate float time shown in the Critical Path Method schedules to advance their own interests. Hence, the

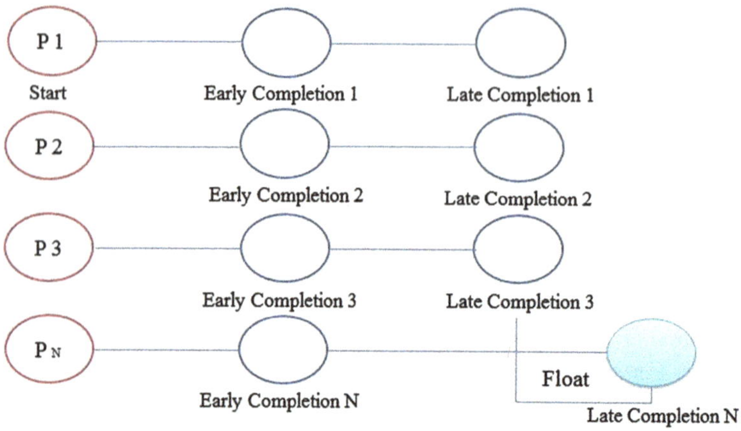


**Fig. 3.9** Float for a program

importance of understanding, analyzing, and utilizing float time properly in a Critical Path Method.

While a degree of float may exist in the activities of a particular Crtical Path Method, the program manager will always seek to minimize the float by reducing its resources appropriately; the program manager will always seek optimum utilization of the available resources. In the circumstances of a certain program, that would mean the use of reduced workforce rather than intermittent use of the workforce. Consequently, bearing in mind that the program manager has no obligation to accelerate the works, it appears that the most likely causes of delay will be such events that prevent a contractor on a project in a large program from working entirely or from working efficiently on activities that increase the productivity.

In the event that the early times of activities are not equal to the late times computed from the Forward Pass and Backward Pass, such activities do not have to start or finish with the early start or early finish times for the program to come in under or on schedule. However, these activities must begin and end with the late start or finish times. The difference between the early and the late start and the finish times indicates the maximum time the activities can be delayed without hindering the program completion.

Several types of float are recognized, including total float, free float, independent float, and interfering float. Figure 3.9 shows an example of float for a program.

## Total Float and Free Float

Total float for an activity is the difference between the early start date and the respective late start date, or between an early finish date and the respective late finish date. In other words, total float for an activity corresponds to the concept of making available to the activity the greatest amount of available float time without jeopardizing the project duration.

Total float, therefore, designates the number of time units belonging to all activities on the same path that can be delayed without extending the program completion time. To simplify, it is the amount of time by which the activities can be delayed without delaying the completion date of the entire program. Total float is an attribute of a network path and does not belong to any one specific activity along that path. Using total float for any single activity on a path thus will reduce the total float times for the following activities. The lower the value of total float, the less flexibility the activity will have for timely program completion.

Free float is the difference between an activity's early finish time and the earliest start time for any succeeding activities. The free float of an activity is the amount of time by which that activity can be delayed without affecting the early start of any following activities or other activities in the network. Unlike total float, free float is the property of an activity and is not shared with any other activities in the network.

Free float is calculated by subtracting an activity's early finish time from the early start time of the next activity. It represents the flexibility that the program manager possesses regarding the start and the finish dates before an effect is felt on other activities in the network. Free float for an activity corresponds to the concept of making available to the activity only that amount of available time that does not interfere with subsequent activities.

Thus, if an arrow notation activity $A_{ij}$ of duration $d_{ij}$ is considered joining nodes $i$ and $j$ and $T(E, ij)$ and $T(L, ij)$ are earliest and latest start node time symbols, then the following float definitions can be made in Eqs. 3.6 and 3.7:

$$\text{Total Float } \mathrm{TF}_{ij} = T(L,j) - [T(E,i) + d_{ij}] \tag{3.6}$$

$$\text{Free Float } \mathrm{FF}_{ij} = T(E,j) - [T(E,i) + d_{ij}]. \tag{3.7}$$

## Interfering Float

That part of the total float which causes the reduction in the float of the successor activities is called **interfering float**. It indicates the portion of the activity float that cannot be consumed without affecting adversely the float of the subsequent activity or activities.

Interfering float = latest finish time of an activity in question minus earliest start of the following activity or zero, whichever is larger in below Eq. 3.8.

$$\begin{aligned}\text{Interfering Float} &= \mathrm{TF}_{ij} - \mathrm{FF}_{ij} \\ &= T(L,j) - T(E,j)\end{aligned} \tag{3.8}$$

## Independent Float

The independent float of an activity is the amount of float time that can be used without affecting either the preceding or succeeding events. It represents the amount of float time available for an activity when its preceding activities are completed at their latest and its succeeding activities have started at their earliest time, leaving the minimum time available for the performance of the activity. Any excess of this minimum time over the duration of the activity is termed the independent float associated with it.

Independent float = earliest start time for the following activity minus latest finish for the proceeding activity minus duration of the present activity.

The use of interfering float by, and during, an activity indicates that subsequent activities are affected in that they can no longer utilize all their previously available float (Householder and Rutland 1990).

## Float Ownership

The float is the amount of time by which an activity or group of activities may be shifted in time without causing delay to a contract completion date. The ownership of the float, which may ultimately determine entitlement to an extension of time as a consequence of the employer delay or other disruptions during the duration of a program, should be adequately addressed in the contract. The extension of time should only be granted to the extent that the employer delay is predicted at the time of the employer risk event to reduce to below zero the total float on the activity paths, and only to the extent that the activity paths are critical to the actual completion date at the time the employer risk event occurs. A contractor should not be automatically entitled to an extension of time merely because an employer delay to progress takes away the contractor's float for a particular activity. The employer delay should only result in an extension of time if it is predicted to reduce the total float on the activity paths affected by the delay.

The float can be divided into the float as it relates to time and the float as it relates to compensation. Whereas the contractor traditionally takes the view that the float belongs to them, to be used as they see fit, the float is a project resource, to be used when the project needs it. In order to determine the ownership of float, the following steps to be taken should clearly:

1. Determine what activities are affected.
2. Calculate the event duration from all affected activities by reference to the last updating.
3. Determine the status of the activities that are affected at the time the variation is issued or when the delay occurs.
4. Create a detailed analysis of the sequence of activities necessary to satisfy the change requirements, or which identify the delay.

If the contractor needs a time buffer for his own use, this should be included as a time contingency in the baseline program (Householder and Rutland 1990).

In a resource-restrained schedule, the concept of float breaks down and quite often the concept of a critical path breaks down. Since almost all construction projects are resource constrained, at least to some extent, this becomes a source of major problems. The classic legal question relates to the ownership of the float is found in Ascon Contracting Ltd v Alfred McAlpine Construction Isle of Man Ltd (1999). In this case, the judge issued a dictum regarding float in a program and its ownership. McAlpine argued that its program for the main contract works contained a float of five weeks. It argued that it had discretion as to which the subcontractor might benefit from this float and accordingly that it might disregard such float in assessing the delay for which Ascon would be held responsible. Judge Hicks held, obiter, that such an argument was misconceived. Not having suffered any loss, the main contractor cannot recover from its subcontractors a hypothetical loss it would have suffered had the float not existed. The issues in any claim against the subcontractor in such circumstances remained simply breach, loss, and causation. It is difficult to claim ownership of something that may not exist or that has not been quantified properly.

## Methods of Calculating Delays

Delays have been found to be the most cited source of disputes and the most costly cause of problems on construction projects in many contractual regimes. Given this state of affairs, it is also noticeable that the cases that have come before the courts, where time disputes involving delay and causation issues are central to the proceedings, rarely involve the use of programming techniques as a method of reliable analysis. Therefore, the courts are increasingly demanding clearer explanations of cause and effect and, in complex construction projects, detailed time impact analysis (Knoke 1995).

The Traditional Forms of Contract make little contractual provisions to integrate the programming of activities into structural obligations. Delay and disruption issues that ought to be managed within the contract all too often become disputes that have to be decided by third parties such as adjudicators, arbitrators, judges, and dispute review boards only after the delays have occurred and disputes have arisen.

Program scheduling consists of updating current and target schedules for existing projects within the program, and developing breakdown structures, milestones, target schedules, and cost-loaded schedules. The resolution of disputes on large construction and engineering programs increasingly involves the use of delay analysis techniques to assist in identifying the cause of critical delay to a program and computing claims for lost productivity. While the industry is becoming more and more familiar with the use of the tools and techniques employed in the process of delay analysis, unfortunately at present, there is very little common agreement upon their correct application and understanding.

Delay, which results in too many programs finishing late and over budget, is often supplemented by enormous claims for compensation or liquidated damages. A delay can be the time during which some part of the construction program has been extended beyond what was originally planned, due to an unanticipated circumstance or circumstances, or it can also be an incident that affects the performance of a particular activity without affecting the program completion. It is possible for a delay not to extend the completion date, but nevertheless to increase the cost to complete particular activities and, therefore, to have the potential for fueling delay claims (Knoke 1995).

As the complexity of programs and the requirements for scheduling have increased, which in turn have augmented the opportunities for delay to the various activities which have been scheduled and are necessary for the completion of the program. In fact, even determining whether completion of the total program or a phase of the program has been delayed can be a difficult analytical task. Since delay usually leads to cost increases, there is a need to correctly determine the allocation of delays as well as the causes and the responsibility of the parties. With this allocation, there can be a technically sound foundation for an acceptable resolution of attributing the costs of delays. With the increasing use of the critical path method, the process of sorting and recognizing the varying situations is facilitated.

In formulating and agreeing a critical path method, particularly where the contractor is responsible for scheduling together the activities, a prudent contractor will normally make some allowance in the critical path for delays that might occur for which he would not be entitled to extensions of time (shortage of labor resources, late delivery of materials, contingency time to allow for unexpected problems encountered in the execution of the works such as specialist subcontractors, overseas orders...).

The Critical Path Method can be used to minimize potential time-related claims, to justify actual claims, and to assist in negotiating timely solutions of both in and out of court settlements. In order to show that an event was not on the critical path, the defendant has to argue that the claimant's version of the critical path is incorrect and must prove on the balance of probabilities that the critical path has shifted elsewhere. Of course, it will also be appropriate to investigate whether, on the facts, the event actually caused a delay and whether it had any consequential effects, as well as looking for other events that may have driven progress at that time and which were therefore the true cause of delay (Palles-Clarke 2003).

There are two schools of thought or methods on how the extension of time should be calculated where an extension of time is granted during a period of culpable delay, which is a delay wholly the responsibility of the contractor.

The first method, described as the gross method, referred to in Fig. 3.10, has been preferred by many academics and some commentators and propounds that if an extension of time is granted because of an event arising during a period of culpable delay, then the extension of time must begin to run from the date the event occurred. This means that the program manager must establish a new completion date for the program which adds the extension of time from the date of the

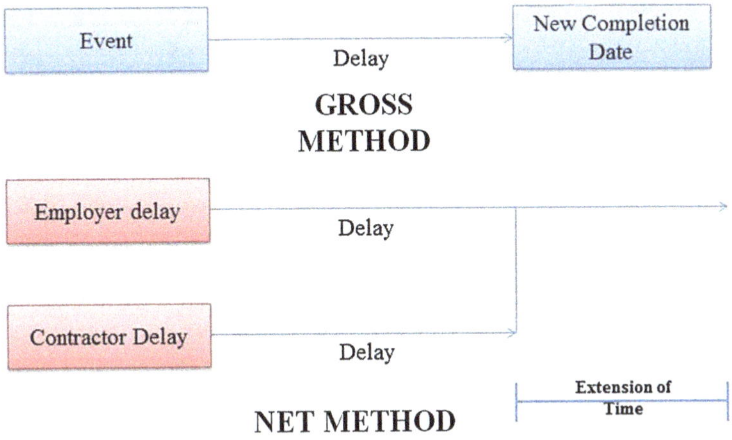


**Fig. 3.10** Gross and net methods

instruction, thus denying the employer liquidated damages up to the new completion date. Naturally, many employers found this to be unfair.

The second method, known as the net method (see Fig. 3.10), simply dictates that the contractor is only entitled to an extension of time equal to the time required to carry out the additional work. Effectively, this means that if the contractor is six months in delay and is delayed by one further month due to a relevant event, the completion date would be extended from the original completion date to a month later, still leaving the contractor with five months of culpable delay and the threat of liquidated damages. Some contractors would consider this to be unfair as the employer may be directly responsible for the late relevant events, e.g., issuing instructions for extra work (Pickavance 2005).

## Other Aspects of Time-Related Issues

The more the number of activities in a program, the greater will be the number of logic links between them and so greater will be the number of assumptions involved in completing the model. Hence, when carrying out retrospective delay analysis using critical path analysis, large programs with hundreds and thousands of activities will produce unreliable results. This is because the analyst has made hundreds of assumptions in preparing the program and, when considering the impact of an event, would probably make many adjustments to a program if faced with a potential delay, especially if the delay is on the critical path.

It is important to recognize that it is easy to manipulate a Critical Path Method in order to derive the required end result. For example, if a programmer wishes to make a certain section of the work critical, he achieves this by fixing durations of activities or logic links between activities. Equally, if variations have been issued in one part of the works, it is possible to make this element of the program critical. This is another reason why it is worth considering reducing the number of activities on very large programs used in retrospective delay analysis.

The float ownership concept is fundamental to the analysis of the project delay and the allocation of responsibility. Both owner and contractor want access to the float in the schedule because it affords them more flexibility in their decision making and use of resources. However, many contracts do not address this important topic. As a result, neither owners nor contractors have a contractual right to use the float. What is now the generally accepted and sometimes disputed answer is that the program owns the float. Under this interpretation, a party is permitted to delay an activity with positive total float provided that the delay duration does not exceed the total float calculation for that activity and that their use of positive total float occurred prior to anyone else's. The float ownership concept becomes more complex when the project is late and the total float calculation becomes negative.

NEC3 (2005) states clearly that there is no reason why the contractor cannot show on his program a date for planned completion earlier than the completion date, thus including some terminal float in the program. NEC3 core Clause 63.3 provides that "*A delay to the completion date is assessed as the length of time that, due to the compensation event, planned completion is later than planned completion as shown on the accepted program.*" Therefore, as stated in the guidance notes to the contract, any terminal float resulting from an early planned completion date is preserved. The period of delay to the planned program is then added to the completion date to determine the revised completion date from which delay damages will be applicable.

The program manager may choose to reverse the impact of delays by expressly ordering acceleration in order to put the program back on schedule and applying the Critical Path Method analysis to show the cost of such a directive. In other cases, where the owner resists to grant the contractor an extension of time that he is entitled to for an excusable event or where an extension has been granted by the owner, but for a shorter period than the contractor is entitled to, a contractor may feel compelled to accelerate the works in order to overrun the completion date set by the owner, thereby avoiding exposure to liquidated damages. To recover under this theory, the contractor must prove:

1. The owner failed to grant an adequate or any extension of time.
2. The owner made it clear that completion was required within the original contract period.
3. Adequate notice had been given by the contractor to the owner advising that he was treating the owner's actions as constructive acceleration.
4. The proof that there had been the actual insurance of additional costs.

If the contract provides for acceleration, payment should be based on the terms of the contract. If the contract makes no provision, the parties should agree the basis of payment before acceleration is commenced. In the absence of agreement, steps should be taken by either party to have the dispute or difference settled (Driver 1994).

Liquidated damages, however, are essential in assessing damages in construction contracts. They are damages calculated according to a formula set out in the contract. They are designed to save the employer from having to prove the actual damage suffered. A clause in a construction contract providing for liquidated damages for delay has to be closely linked with a clause that provides for an extension of time. This is because it is assumed that if the contractor is to be held liable for liquidated damages, the delay for which damages are to be calculated must be the responsibility of the contractor. If delays caused are the responsibility of the employer, then the contract must provide a mechanism whereby the date for completion can be extended.

If the contract fails to provide for such an extension, or if the architect or engineer fails to administer the extension of time provisions correctly, then the liquidated damages clause may become unworkable, if only because there is no fixed date from which to calculate the delay for which the contractor is responsible.

## Critical Path Method's Dynamic Nature—The Legal Approach

In Henry Boot Construction (UK) Ltd v Malmaison Hotel (Manchester) Ltd (2000), the court clarified that when it is agreed that there are two concurrent causes of delay, one of which is a relevant event and the other is not, then the contractor is entitled to an extension of time for the period of delay caused by the relevant event, notwithstanding the concurrent effect of the other event. Thus, to take a simple example, if no work is possible on a site for a week, not only because of exceptionally inclement weather (a relevant event) but also because the contractor has a shortage of labor (not a relevant event), if the failure to work during that week is likely to delay the works beyond the completion date by one week, and then, if he considers it fair and reasonable to do so, the architect is required to grant an extension of time of one week. The program manager is not precluded from considering the effect of other events when determining whether a relevant event is likely to cause delay to the works beyond completion.

Lord Denning explained, obiter dicta, in the case of Amalgamated Building Contractors Ltd v Waltham Holy Cross (1952) that the power of extending contractual completion dates should apply retrospectively and that common sense required that the method of assessment of such an extension would be what he termed the net method. Lord Denning stated:

> *Take a simple case where the contractors, near the end of the work, have overrun the contract time for six months without legitimate excuse. They cannot get an extension for the period. Now suppose that the works are still uncompleted and a strike occurs and lasts a month. The contractors can get an extension of time for that month. The architect can clearly issue a certificate which will operate retrospectively. He extends the time by one month from the original completion date, and the extended time will obviously be a date which is already past.*

Peak Construction (Liverpool) Ltd v McKinney Foundations (1970) related to a case where the works were suspended due to defective piles for which the contractor was responsible, but the employer caused further delay for which there was no mechanism in the contract for extending time. The result was that it was not possible to fix a new completion under the contract, time was at large, and the employer was denied the right to liquidated damages for any of the delay. The main contract allowed the architect to certify extensions of time for additions to the works, strikes, force majeure, or "*any other unavoidable circumstances.*" These provisions did not permit an extension of time to be granted for the employer's failure to promptly authorize and instruct the investigations and remedial works. Such a failure was clearly avoidable.

In Glenlion Construction Ltd v Guinness Trust (1987), the issue that arose was whether Glenlion was entitled to complete the works before the completion date. Glenlion had submitted a program which showed early completion. It was held that it was self-evident from Clause 21 of *JCT 63* that Glenlion was entitled to complete before the date of completion. This was so whether or not Glenlion produced a program with an earlier date and whether or not the company was required to produce a program. Judge Fox-Andrews QC commented:

> *In regard to claims based on delay, litigious contractors frequently supplied to architects or engineers at an early stage in the work highly optimistic programs showing completion a considerable time ahead of the contract date. These documents are then used (a) to justify allegations that the information or possession has been supplied late and (b) to increase the alleged period of delay, or to make a claim possible where the contract completion date has not in the event been extended.*

While Glenlion was entitled to complete before the contractual completion date, it was held that Guinness was not required to actively cooperate to enable the earlier date to be achieved but was only required not to hinder completion. It is suggested that the situation will, however, be different if the program is incorporated in the contract as a contract document or if the entitlement under the contract is bound up with the program. If the employer does not wish to take possession of the works early, then this needs to be dealt with by amendment of the contract terms so that the contractor can price accordingly. In *Glenlion Construction Ltd v Guinness Trust,* it was also recognized that the concept of float is a main contractor program and its ownership was addressed. It was established in this case that an employer does not have to assist the contractor in his efforts to complete early, but nevertheless he should refrain from doing anything that would deliberately hinder early completion.

In Balfour Building Ltd v Chestermount Properties Ltd (1993), the court said that the purpose of the power to grant an extension of time was to fix the period of time by which the period available for completion ought to be extended having regard to the incidence of relevant events. The completion date, as adjusted, was not the date by which the contractor ought to have achieved practical completion, but the end of the total number of working days, starting from the date of possession, within which the contractor ought fairly and reasonably to have completed the works. On this footing, where a relevant event arose after the date for completion and during a period in which the contractor was in culpable delay, the contractor would only become entitled to a "net" extension of time corresponding to the specific number of days of delay occasioned by the relevant event. In other words, the occurrence of the new delaying event would not let the contractor off the hook for its own culpable delays.

With regard to the questions raised in the Balfour case, Judge Toulmin concluded that an extension of time for the completion of the works may be granted in respect of a relevant event occurring during the period of culpable delay. However, he refused to follow precisely the guidance in Balfour to determine the net effect of delays occurring after the date for completion.

It is worth noting that the case of *Balfour Building v Chestermount* is one which establishes a very important principle in considering any entitlement to extensions of time and which represents a commonsense approach to the issue, namely that Clause 25.3 of JCT 80 was wide enough to include relevant events that occurred after as well as before any previously fixed completion date, and also that when considering delay events that occur in a period of culpable delay, the contractor will be entitled to an extension of time on the net method of extension, which allows the incremental time lost for the new event to be added back to the previously extended completion date, rather than allowing an extension of time to the date on which the late instructed work is completed. In other words, the gross method of extension was disapproved. As Nicholas Carnell has observed:

> *In the context of the power to review it was contended that on a proper construction of clause 25.3.3, the power to review could only be exercised to grant the new completion date at a future date. Mr. Justice Colman rejected this proposition, and held that the duty was to review the net extension to which the contractor was due, and that this could in many instances result in the completion date being fixed at a date prior to the date on which the review had taken place.*

In John Barker Construction Ltd v London Portman Hotel Ltd (1996), John Barker were building contractors carrying out refurbishment of works to the London Portman Hotel. The contract was the JCT 80 form with quantities. The contract provided for the completion of floors 9–11 on July 16, 1994; floors 5–8 by July 29, 1994; and floors 2–4 by August 14, 1994. Clause 24 provided that liquidated damages would be paid at £30,000 per week for each section of the contract that was not completed by the specified date. Delays occurred and it was apparent to all concerned that John Barker was entitled to extensions of time. After negotiations, it was agreed that the work would be accelerated so that all the work would be completed by August 14, 1994, and John Barker would receive additional payment.

After the acceleration agreement, there were further delays and further instructions from the architect. One of the issues that arose was the effect of the acceleration agreement on the sectional completion provisions of the contract in relation to liquidated damages.

John Barker argued that the effect of the acceleration agreement was to dispense with all the provisions of the sectional agreement supplement, including the provisions for liquidated damages. It was argued that the substitution of a single date was logically inconsistent with such provisions having continuing contractual force. This was not accepted. It was common ground that at the time of the acceleration agreement, no party raised the question of abandoning the liquidated damages provisions. It was held that it was neither intended by the parties nor logically necessary that the liquidated damages would no longer apply. It was held that the provisions of the sectional completion supplement regarding liquidated damages were capable of continuing to have contractual force by merely substituting the new date of August 26, 1994, for the completion of each section.

Mr. Recorder Toulson QC emphasized that in exercising his duty, under Clause 25 of the JCT Standard Form of Contract, the architect or contract administrator must undertake a logical analysis in a methodological way of the impact of the relevant events on the contractor's program. The application of an impressionistic rather than a calculated and rational assessment is not sufficient. A detailed analysis would lead to the real point behind the case, which is that in practice, Mr. Recorder Toulson was calling for a significant extension of time assessments to be done by some Critical Path Methods.

Henry Boot Construction (UK) Ltd v Malmaison Hotel (Manchester) Ltd (2000) addressed directly how delays caused by a contractor affected its entitlement to extensions of time for delays that were the responsibility of the employer. Malmaison engaged Boot to construct a new hotel in Piccadilly, Manchester. Practical completion was fixed for November 21, 1997, but was not achieved until March 13, 1998. The architect extended the time to January 6, 1998. On the strength of those certificates, Malmaison deducted liquidated damages totaling £250,000 from Boot's account. Boot claimed further extensions of time in respect of a number of alleged relevant events. However, for tactical reasons, it gave notice of arbitration in respect of only two, which it used to claim extensions through to practical completion. If it succeeded in the arbitration, the liquidated damages would have to be repaid. It would, then, also be in a position to claim time-related costs for the overrun. Malmaison denied that the two alleged relevant events caused delay. It went further by pleading a host of other matters which, if proved, would demonstrate that Boot itself was the cause of the overrun.

The main employer stance was: (1) how can the main contractor be entitled to an extension of time? (2) when and if we did delay him, he was in delay himself; and (3) the main contractor was not in a position to undertake the scheduled work in any event. The logical step to be taken from Malmaison is to conclude that only activities on the critical path have an effect on the completion date.

Thus, in deciding whether to grant an extension, the architect must take due cognizance of whether the activity so delayed is or has become, due to the delaying event, on the critical path. This is where extension of time can go wrong. The basic factual matrix is often not sufficiently established or understood when the claim is prepared, with the result that the proper context in which an event occurred is often when it is alleged that an event has delayed the completion date. The above indicates that the courts are concerned primarily with what has actually happened rather than considering any sort of analysis that may be based on some speculation about what would or could have been the effect at the time the event occurred. Judges and arbitrators are really looking to get into the facts in order to find out what really happened on the site and to identify the real causes of delay. It is important to recognize that there is a distinction between what really caused delay and a contractor's entitlement to an extension of time in accordance with the terms of the contract.

In Royal Brompton Hospital NHS Trust v Hammond & Others (No 7) (2001), the court considered allegations of negligence by the claimant's professional advisors in granting extensions of time under a building contract. The court held that principles of professional negligence applied to the claims against each of the defendants. The architects were being sued on the basis that they had been far too generous in granting extensions of time. One of their many responses was to say that it was perfectly appropriate to give extensions of time where there were concurrent culpable and non-culpable causes of delay. However, the court was not directly concerned with the issue of whether extensions of time should or should not be granted where there is true concurrency. Judge Seymour postulated different types of concurrency and suggested that Mr. Justice Dyson was referring to a type of concurrency, which he called a "true concurrency." This was supposed to be concurrency where, work having already been delayed for, say, shortage of labor, an event occurs which is a relevant event and which, had the contractor not been delayed, would have caused him to be delayed but which in fact, by reason of the existing delay, made no difference.

This is clearly difficult to reconcile with the decision in Balfour Building v Chestermount (1993) and Henry Boot v Malmaison (2000). Judge Seymour has made the following comment:

> *There may well be circumstances where a relevant event has an impact on the progress of the works during a period of culpable delay if where that event would have been wholly avoided had the contractor completed the works by the previously fixed completion date. For example, a storm which floods the site during a period of culpable delay and interrupts progress of the works would have been avoided altogether if the contractor had not overrun the completion date. In such a case it is hard to see that it would be fair and reasonable to postpone the completion date to extend the contractor's time.*

Looking at this example, perhaps, this would appear not to be so unreasonable and may indeed generally reflect the JCT fair and reasonable requirements for the architect or employer's agent to consider an extension of time. Unfortunately, this interpretation may limit the scope of the decision in Malmaison and may in fact encourage the dominant approach.

The John Barker (1996) case was considered in the recent case of Balfour Beatty v The Mayor and Burgess of the London Borough of Lambeth (2002) which concerned a challenge to the enforcement of an adjudicator's decision on the basis that in reaching his decision, the adjudicator had failed to act fairly and had breached the rules of natural justice by preparing his own programming analysis in the absence of one from the referring party, but had done so without giving the responding party an opportunity to comment on the methodology or his approach. In reaching his decision, His Honour Judge Humphrey Lloyd QC observed that:

> *Despite the fact that the dispute concerned a multi-million pound refurbishment contract, no attempt was made to provide any critical path. The work itself was no more complex than any other projects where a Critical Path Method is routinely established and maintained. It seems that BB had not prepared or maintained a proper program during the execution of the works. ... [I]n the context of a dispute about the time for completion a logical analysis includes the logic required for the establishment of CPN [Critical path network].*

In Balfour Beatty, the court also emphasized that the purpose of the power to grant an extension of time was to fix the period of time by which the period available for completion ought to be extended having regard to the incidence of relevant events. The completion date, as adjusted, was not the date by which the contractor ought to have achieved practical completion, but the end of the total number of working days, starting from the date of possession, within which the contractor ought fairly and reasonably to have completed the works. On this footing, where a relevant event arose after the date for completion and during a period in which the contractor was in culpable delay, the contractor would only become entitled to a net extension of time corresponding to the specific number of days of delay occasioned by the relevant event. In other words, the occurrence of the new delaying event would not let the contractor off the hook for its own culpable delays.

These two cases were reconsidered in the case of Motherwell Bridge Construction Ltd (T/A Motherwell Storage Tanks) v (1) Micafil Vakkuumtechnik Ag (2) Micafil Ag (2002). Micafil was engaged by BICC as main contractor for the construction of an autoclave, a large steel vessel with an internal volume of 650 $m^3$. The vessel was to be used in the manufacture of high-quality power cables. Micafil undertook responsibility for the design of the vessel and subcontracted its construction to Motherwell Bridge. During construction, Motherwell Bridge raised many technical queries and there were a number of significant design changes issued by Micafil. There were two major formal amendments to the contract. Delays occurred and Micafil deducted liquidated damages. Motherwell Bridge in turn claimed extensions of time to extinguish the claim for liquidated damages. His Honour Judge Toulmin QC commented:

> *Crucial questions are (a) is the delay in the critical path? and if so, (b) is it caused by Motherwell Bridge? If the answer to the first question is yes and the second question is no, then I must assess how many additional working days should be included.*

He added:

> *Other delays caused by Motherwell Bridge (if proved) are not relevant, since the overall time allowed for under the contract may well include the need to carry out remedial works or other contingencies. These are not Relevant Events, since the court is concerned with considering extensions of time within which the contract must be completed.*

Judge Toulmin went on to say that the approach must always be tested against an overall requirement that the result accords with common sense and fairness.

In order to show that an event was not on the critical path, the defendant has to argue that the claimant's version of the critical path is incorrect and must prove on the balance of probabilities that the critical path went elsewhere. Of course, it will also be appropriate to investigate whether, on the facts, the event actually caused a delay and whether it had any consequential effects, as well as looking for other events that may have driven progress at that time and which were therefore the true cause of delay.

In Multiplex Constructions (UK) Ltd v Honeywell Control Systems Ltd (2007), Honeywell, responsible for the security and communication installations at Wembley Stadium, contended that the project was so mismanaged by Multiplex, the main contractor, that it was no longer bound by the terms of its subcontract to complete its works by a fixed date. Honeywell argued that time had become at large and that its obligation to complete its subcontract works within a specified time had fallen away, only to be replaced by an obligation to complete the works within a reasonable time. Honeywell argued that time had become at large on four different grounds:

1. The issue of further programs as a direction under Clause 4.2 meant that there was a delay to the finishing of the works and no corresponding relevant event existed in Clause 11 for Honeywell to apply for an extension of time under the "sweep up" relevant event clause, which was couched in terms so that it only applied to unlawful actions by Multiplex.
2. Honeywell argued that Multiplex had simply failed to operate the extension of time machinery in the subcontract and, in the alternative, that machinery had broken down.
3. The condition precedent barred Honeywell from an extension of time and this was not allowed.
4. The compromise agreement struck between Multiplex and Wembley National Stadium Limited superseded the extension of time mechanism in the main contract.

Multiplex commenced a preemptive action in the Court to secure a decision that the contract mechanisms for fixing the date for completion remained operative. Honeywell founded this argument on an old concept in the general law, the prevention principle. The essence of the prevention principle is that the employer cannot hold the contractor to a specified completion date if the employer has, by an act or omission, prevented the contractor from completing by that date. Instead, time becomes at large and the obligation to complete by the specified date is replaced by an implied obligation to complete within a reasonable time. It is in order to avoid the prevention principle that many construction contracts and subcontracts include provisions for extension of time. The 1970 case of Peak Construction foundations continues to underpin the point where the extension of time provisions did not embrace causes of delay which were the employer's own fault. Clause 11 clearly gave Multiplex the right to grant extensions of time for "acts of prevention" and a direction permitted by the contract, namely issuing a revised program under Clause 4.2, was an act of prevention for which an extension could be granted.

Honeywell argued that Multiplex had made the extension of time machinery inoperable by its failure to provide adequate programming data to the subcontractor. Throughout the works, Honeywell had been obliged to work to short-term programs and it argued that it had never been able to plan its works or to identify the impact of delay events upon the critical path for completion. This in turn meant that it could never effectively comply with the subcontract requirement for notices and particulars of delay which the subcontract made a condition of Honeywell's rights to extensions of time.

On a matter of construction, His Honour Justice Jackson found that the sweep up provisions of Clause 11 entitled an application for an extension of time to be made in the event that the works were varied pursuant to Clause 4.2. This is because an act of prevention may be a legitimate act, an act of prevention being referred to in the pertinent part of Clause 11, and consequently, time was not at large on this ground. The Technology and Construction Court also found that the extension of time machinery had been operated and had not broken down. Honeywell was only required to do its best in supplying notices and information and Honeywell's own evidence stated it had done just that. It was well-established law that a party cannot insist upon the performance of an obligation which he has prevented a promisor from performing. However, in this case, the construction contract permitted Honeywell to obtain an extension of time to circumvent any such proceeding possibilities and thus, time was not at large.

## US Construction Case Law Pertinent to Extension of Time and Delays

In a notorious case, Natkin & Co v George A. Fuller Co. (1972), one of the findings adopted by the judge in order to reach his decision was that "*the Critical Path Plan may become obsolete unless it is kept current.*" Other findings noted were that the subcontractor Natkin objected to the use of charts in lieu of Critical Path Method updates and that the Critical Path Method required updating at different phases of the project. The judge emphasized that "a *Critical Path Method schedule's usefulness as a barometer for measuring time extensions and delay damages is necessarily circumscribed by the extent to which it is employed in an accurate and consistent manner to comport with the events actually occurring on the job.*" Furthermore, updating the Critical Path Method during the life of the project is incremental; not doing so can make the schedule redundant.

In Fortec Constructors v United States (1985), Fortec, the contractor, sought time extensions, extra costs, and liquidated damages for a government project to build a fuel maintenance facility on an Air Force base. The Claims Court rejected a Critical Path Method prepared by the US government to show that the extra work was not on the critical path, and, therefore, the contractor was not entitled to any time extensions. The reason given was that "*...if the Critical Path Method is to be used to evaluate delay on the project, it must be kept current and must reflect delays as they occur...*".

In the Preston-Brady Co. (1991) case, the Court of Appeal found that in addition to establishing the schedule updating process as an essential requirement in any construction project, the owners failed in approving schedules and issuing change orders and corresponding time extensions in a timely manner.

The importance of the issues of delay and disruption was emphasized in the case of R P Wallace v The United States (2005), where the judge residing described the development of law in the area of delay and disruption in the USA as akin to development of other important laws in the nineteenth and twentieth centuries such as the law of tort or negligence.

In John Driggs Company (2005), the contractor sought additional compensation and extensions in contract time for multiple events that occurred early in the contract. The Board commented that a common thread running through all of these alleged "delays" is that Driggs did not complete these particular tasks on the originally planned and scheduled date and therefore, when a significant owner-caused construction delay occurs, the contractor is not necessarily required to conduct all of his other construction activities exactly according to his predelay schedule and without regard to the changed circumstances resulting from the delay. The occurrence of a significant delay generally will affect related work, as the contractor's attention turns to overcoming the delay rather than slavishly following its now meaningless schedule.

The *Driggs* decision appears to affirm the notion of reallocation of resources in observance of another, more critical delay and the contractual right to utilize float (even if it is negative) made available by another party's critical path.

## Standard Forms of Contract and Scheduling

In order to prove a delay claim, it is necessary for the claimant to satisfy the requirements of the particular contract under which he is operating. This is likely to require that firstly, a particular event is of a type that gives rise to an entitlement under the contract for an extension of time; secondly that the event has occurred and its extent is quantified; and thirdly, that its consequences are identified by establishing that the event caused delay to the completion of the project.

In the context of construction programs, however, this is often very difficult to do, and the extent to which this is possible is usually a function of the quantity and quality of the records held by the contractor about what actually happened. It is a generally accepted principle of risk management that those who are most able to manage a particular risk should bear that risk, since "*No construction project is free of risk. Risk can be managed, minimized, shared, transferred or accepted. It cannot be ignored*" as stated by Sir Michael Latham in his widely publicized Final Report of the Government/Industry Review of Procurement and Contractual Arrangements in the Construction Industry entitled *Constructing the Team* (1994). However, in the Standard Forms of Contract for construction and engineering, that principle becomes absolute. Once the contractor is appointed and the work commences on site, all the risk of change or interference that is left in the hands of the employer cannot be managed because all the information needed to manage that risk is left in the hands of the contractor.

The Association of Consultant Architects contract PPC (2000) does not even require a program. JCT 98 (1998) requires a program but does not require it to be updated to save for the change in completion date on the award of an extension of time. No standard forms require the contractor to provide contemporaneous records of progress actually achieved. Of course, the contractor is required to use the tools he has to manage the employer's risk and to "overcome and avoid unnecessary delay howsoever caused," but, if he fails to do so, he is to be compensated for any loss he suffers and given more time to complete. See, for example, JCT 98 (1998), Clause 25.3.4.1, and ACA contract PPC (2000), Clause 18.3. History has shown the possibility of the contractor managing the employer's risks to be a pious hope and it is perhaps not surprising that under the current standard forms of contract, there is little impetus for the contractor to do so.

Standard forms of contract have a tradition of weakness and "forgetfulness" as to program obligations. The most continuing example is the continuing failure of JCT, ICE, and Fédération International des Ingénieurs-Conseils (FIDIC) properly to integrate the program as a full contractual mechanism. Rather, it remains either incidental or merely a management information tool. Most standard forms of construction contract require the architect or engineer to deal with the contractor's claims for extension of time when actual delay has occurred to the works and also when a prospective delay event is notified. An example of this approach is found in the JCT Standard Building Contract 2005 at Clause 2.27:

> *if and whenever it becomes reasonably apparent that the progress of the Works or any Section is being or is likely to be delayed the contractor shall forthwith give written notice…*

Under NEC3 (2005), the program is a much more important tool for the administration of progress and delay. Core Clause 31.2 sets out the requirements of each program that the contractor submits, including the requirement to show the completion date (as stated in the contract data or as extended by the contract), planned completion (which may be a different date), and any provisions for float. The intention, in accordance with core Clause 3, is that the program is revised monthly or otherwise when the contractor chooses or the project manager instructs. The program is required to reflect the state of affairs at that time, including the effect of any compensation events for which an allowance for delay has been determined, as well as other delays to date. Problems arise where the contractor has not submitted an updated program or where the project manager does not accept the latest program, in which case the project manager assesses the compensation event using his own assessment of the remaining program.

## The Critical Path Method and Standard Forms of Contract

The main parties involved in the contract are often reluctant to allow the initial tender program to have the status of a contract document. This reluctance often stems from the traditional hierarchy of the construction industry, and from the in-built fear that the parties involved lack the expertise in the programming techniques and, furthermore, the practical application of such programming techniques in order to achieve the target time and eventual cost.

Because of the nature of construction, the standard forms of construction contracts tend to have a certain amount of flexibility built in that enables the work being carried out to be varied as it proceeds, thus enabling both parties to fulfill their obligations in accordance with the terms under which they are contracting.

In terms of the program use during the post-contract period, although the program will provide a support in terms of better understanding of the impact of delays, it will not necessarily establish the extent of delay unless more sophisticated forms of analysis are adopted and are used on an on-going basis throughout the program so that the program becomes literally a working and living program. This is because if delay allegations are to be shown effectively by the contractor and considered properly by the architect or engineer, it will be found that in most situations, a simple bar chart will not suffice and some better means of indicating quantity output or physical progress, as well as the passage of time, is essential. The purpose of the Critical Path Method in connection with claims is to identify the causes of delays, the dates of onset and cessation, and also, as far as possible, the effects, both immediate and consequential, on the various operations.

New standard forms of contract have emerged, in particular the New Engineering Contract where the program and time are given a more dynamic status. There is also a gradual recognition at long last that the planner should be an integral member of the employer's design team and therefore provide the employer with the ability and confidence to accept the tender program as a contract document. Parties to a construction contract often end up in disputes because the form of contract under which the project is procured, while providing a mechanism for dealing with delay and disruption, does not guide the parties in terms of how the resulting claims should be prepared and managed, and what procedures should be followed during the currency of a contract to enable solutions to be arrived at that sufficiently prevent the parties from falling into dispute.

## The JCT Standard Form of Building Contract, 1998 Edition (JCT 98) and 2005 Edition (JCT 2005)

JCT 98, and similarly JCT 2005, is administered by an architect who is appointed by the employer and acts as his agent, but is required to act fairly and reasonably toward the contractor. The JCT considers the issue of extension of time and reimbursement of loss and/or expense at Clause 25 and Clause 26. Clause 25 sets out the grounds that entitle the contractor to an extension of time. The purpose of this clause is to allow the period of the contract to be extended by adjusting the date for completion beyond the original date of completion.

It is the contractor's duty to give written notice stating the cause or causes of delay when it becomes reasonably apparent that the progress of the works "is being or is likely to be delayed." Such a notice is a condition precedent to the architect's consideration of granting an extension of time under Clause 25.3 prior to practical completion. The contractor must also identify which of the relevant events he considers appropriate.

Clause 25.4 provides a list of 18 categories of delay, which are known as Relevant Events that give rise to an entitlement to apply for an extension of time in the event that they occur. The principal steps set out in the clause are as follows:

1. When it becomes apparent that the progress of the works is being or is likely to be delayed, then the contractor shall notify the architect of the cause of delay and identify whether in his opinion it is a Relevant Event.
2. The contractor is required to provide with the notice, or as soon as possible after the notice, particulars of the expected effect of the event and an estimate if any of the expected delay to the completion of the works beyond the completion date.
3. Upon the receipt of the notice and any further particulars, the architect is required to decide whether in his opinion any of the events notified are Relevant Events and whether as a result of such events the works are likely to be delayed beyond the completion date. If he so decides, then he is then required to give an extension of time to the contractor in writing as he then estimates to be fair and reasonable.

This whole procedure is quite clearly intended to be carried out before completion date since reference is clearly made to the contractor notifying the expected effect of the event, not necessarily the actual effect of the event, and the architect is required to make a fair and reasonable estimate of the likely effect on the completion date, not wait until the completion date has arisen to determine the precise effect.

Events that are outside the control of either party, also known as "neutral events," shall give rise to an entitlement of extension of time for completion but not for the reimbursement of loss and/or expense occurring as a result of such an act or default arising. Clause 25 essentially comprises six parts that relate directly to extension of time:

1. The requirement for the contractor to give timely notice of delay and to give further details to assist the architect in establishing any extension of time.
2. The requirement for the architect in certain circumstances to grant extensions of time during the construction periods or to notify the contractor that an extension of time is in his view not justified.
3. The requirement or permission for the architect to review the completion date after completion.

4. The effect of omissions upon extensions of time.
5. The list of events which entitle the architect to consider granting an extension of time—known as Relevant Events.
6. The general obligations imposed upon the architect and the contractor in relation to their conduct.

Under Clause 25.3.1.2, there is provision whereby the contractor is required to submit two copies of his master program for the execution of the works. In addition, the contractor is obliged to amend the program every time the architect makes a decision, under Clause 25.3, to fix a new date for completing the works.

Where, for example, there are three causes of excusable, time-compensatory delay and if each one had delaying effects of three, two, and four weeks, respectively, but they overlapped and had the same start date, the architect would not grant a nine-week extension, but if, say, the first two delays were completely absorbed in the longest, he would grant a total extension of four weeks. In complying with the notice provisions under Clause 25.2.2.2, the contractor is required to state the effect of each delay as if it had been the only one and it will then fall upon the architect to determine their cumulative effect and grant an extension of time accordingly.

## NEC3 Engineering and Construction Contract

The New Engineering Contract (NEC3) gives programming and time management more dynamic status plus the gradual recognition at long last that the planner should be an integral member of the employer's design team. It, therefore, provides the employer with the ability and confidence to accept the tender program as a contract document. The NEC3 is administered by a project manager, and under this form, the program assumes greater importance than under the other forms. It introduces specific terms and definitions of important factors arising from contracts such as *Accepted Program*, *Activity Schedule*, *Completion Date*, *Acceleration for Early Completion*, and *Terminal Float*.

The UK Olympic Delivery Authority (ODA) chose the NEC3 suite of contracts to procure all fixed assets and infrastructure programs for the London 2012 Olympic Games. Around £2.4 billion of new facilities needed to be designed and built for the games, and this figure approached £4 billion due to cost escalation.

There was no separate extension of time provision, but instead, the contract provided for compensation events in accordance with core Clause 60.1, which lists 18 types of compensation event. Under this clause, the project manager may notify the contractor of a compensation event at the time of the event, where the event arises from the project manager, and request the contractor to provide a quotation for the compensation event.

The procedure for notifying, assessing, and implementing compensation events is as follows:

1. The contractor otherwise notifies the project manager of any compensation event which has happened, or which he expects to happen, within two weeks of becoming aware of any event.
2. In order for the prices and/or the completion date to be changed, the project manager must decide that the event has not arisen from any default of the contractor, that it is a compensation event, and that it will have an effect on actual cost completion, in which case he requests the contractor to submit a quotation.
3. A delay to the completion date is assessed by the project manager as the length of time that, due to the compensation event, planned completion is later than planned completion shown on the accepted program.

The combined provisions of Clause 31 and Clause 32 envisage a program that is presented for acceptance at the commencement of the contract and is regularly updated in order to monitor the progress of works. This active project management tool is central to the progress of work and the administration of NEC3.

Hence, NEC3 contracts have "raised the bar" in terms of program management, which in turn has required greater understanding of programming issues by project teams. The most popular NEC form, the Engineering and Construction Contract (ECC), obliges contractors to produce and maintain, and project managers to accept, a progressed "live" program. ECC forces key aspects such as *Float*, *LogicLinks*, *Time Risk Allowance,* and key dates to be understood, planned, monitored, and revised as appropriate. It becomes a key tool to aid parties to make decisions, often as a result of notified early warnings. It provides a thorough audit trail for all parties to assess, in particular, the reasons for any delay at the point at which they occur.

The two biggest issues that are commonly voiced in terms of program are, firstly, that the program manager does not accept or is mute with regard to acceptance of a submitted revised program and, secondly, the program submitted by a contractor has insufficient detail as defined by Clause 31.2 of the contract.

Both scenarios result in the fact that a revised or accepted program is not in place for a period of time, and in some cases for the life of the project. This completely undermines the intent of the contract and the systems it puts in place to ensure "best practice" project management. It is only common sense that both parties would want to understand where the contract actually is at any point in time in terms of financial or time implications.

In terms of the first issue, the contractor's recourse is to notify an early warning if the period for reply is exceeded following a program issue, which should at least prompt a risk-reduction meeting to discuss it. Within the industry, there still appears to be too large a percentage of projects that do not have an accepted program in place, or have an out-of-date program with current works progressing to a significantly different sequence or timescale. It is everyone's responsibility on a project to play their part to ensure this does not happen.

With regard to the second issue, it has been voiced many times before that it is "impossible" for a contractor to comply with Clause 31.2 of the contract. However, it is important to remember the intent of NEC3. It is a stimulus to good management and looks to ensure that the program becomes a tool that the whole project team can use, rather than merely being a retrospective or theoretical high-level reporting tool.

There is in fact nothing within Clause 31.2 that a competent contractor would not or should not consider during part of the tender process in order to ascertain both anticipated cost and duration of a project. It is a case of taking that information and being able to show, monitor, and develop it for the duration of the project.

The two principles on which the NEC3 is based and which impact upon the objective of stimulating good management are, essentially, that collaborative foresight can be applied to mitigate problems and reduce risk and that the clear division of function and responsibility will help accountability and motivate people to play their part. The NEC3 allocates the risks between the parties clearly and simply. A prominent example of the way that the procedures in the NEC3 are designed to stimulate good management is the early warning procedure. This is designed to ensure that the parties are made aware as soon as possible of any event that may delay completion of the works. Problems arise where the contractor has not submitted an updated program or where the project manager does not accept the latest program, in which case the project manager assesses the compensation event using his own assessment of the remaining program.

## ICE Conditions of Contract, Measurement Version, 7th Edition, September 1999 (ICE 7th)

This is a family of standard conditions of contract for civil engineering works, which is at the forefront of best practice and modern procurement methods.

The ICE Conditions of Contract, which have been in use for over fifty years, were designed to standardize the duties of contractors, employers, and engineers and to distribute the risks inherent in civil engineering to those best able to manage them.

ICE 7th is administered by an engineer who is appointed by the employer and acts as his agent, but is required to act fairly toward the contractor. This form of contract provides an entitlement to extensions of time under several of its clauses, but the principal provisions for dealing with extensions of time are set out at Clause 44 as follows:

1. If one of the listed types of delay occurs, then within 28 days after the cause of delay has arisen or as soon thereafter as is reasonable to deliver to the engineer full and detailed particulars in jurisdiction of the period of extension claimed in order that the claim may be investigated at the time (Clause 44(1)).
2. The engineer is then required to make an assessment of the delay having regard to all of the circumstances known to him at the time, whether or not the contractor has made a claim (Clause 44(2)).

3. If the engineer considers that the delays suffered fairly entitle the contractor to an extension of the time for substantial completion of the works, then he grants the extension to the contractor in writing (Clause 44(3)).
4. Not later than 14 days after the due date or extended date for completion, the engineer is then required to make a further interim assessment based on all of the circumstances known to him at the time and whether or not the contractor has made claim (44(4)).

Clause 44 is the focal point of these provisions and is the mechanism whereby the contractor is to notify the engineer of events that he considers may entitle him to an extension of time for the substantial completion of the works.

Clause 44 gives five events that may entitle the contractor to an extension of time. These are variations ordered by the engineer, increased quantities, exceptionally adverse weather conditions, actions or omissions by the employer, and special conditions of any kind which may occur.

The contract requires the engineer to consider all the circumstances known to him at that time and makes an assessment of the delay suffered by the contractor. The contract gives the engineer wide powers to make an assessment of delay even without a claim being submitted by the contractor. The contract also provides for an interim determination of the contractor's time entitlement.

Clause 44(1) states that only events that have an effect upon "substantial completion" entitle the contractor to an extension of time. Therefore, criticality of events is an issue that appears to be a problem with the assessment of a contractor's entitlement as the works proceed. As things change on site, so can the criticality of the events. What starts as a critical event may well not end up being critical.

## Completion, Early Completion, and Acceleration

If, as a result of an employer delay, the contractor is prevented from completing the works by the contractor's planned completion date (being a date earlier than the contract completion date), the contractor should in principle be entitled to be paid the costs directly caused by the employer delay, notwithstanding the employer is aware of the contractor's intention to complete the works prior to the contract completion date, and that intention is realistic and achievable.

If, as a result of a compensable employer risk event that causes delay, the contractor is prevented from completing the works by his own planned completion date (being a date earlier than the contract completion date), the contractor should, in principle, be entitled to be paid the costs associated with such an event. In other words, the float associated with the contractor's planned completion date belongs to him, and the date for completion becomes the date by which the contractor expects to complete the works. This confirms the view that in order to establish that an event has affected the completion date, you must show that it falls on the critical path. Some means by which the critical path can be identified and shown therefore appears to be required. Although the emphasis appears to be on an examination of what actually happened, the passage does not make it clear whether it refers to the critical path at the time of the event or the overall as-built critical path evident at the completion of the project.

In terms of the completion date, a contractor under a JCT 98 form is entitled to program to finish the works before the completion date stated in the contract, but the fact that he has done so does not place any obligation on the architect to produce information by dates earlier than would be necessary for completion by the contract date. In Glenlion Construction v The Guinness Trust (1987), it was established that an employer does not have to assist the contractor in his efforts to complete early, but nevertheless, he should refrain from doing anything that would deliberately hinder early completion.

An employer may choose to reverse the impact of delays by expressly ordering acceleration in order to put the program back on schedule. An analysis of the Critical Path Method shows the cost of such a directive. In other cases, where the owner resists granting the contractor an extension of time that he is entitled to for an excusable event, or where an extension has been granted by the owner but for a shorter time than the contractor is entitled to, a contractor may feel compelled to accelerate the works in order to overrun the completion date set by the owner, thereby avoiding exposure to liquidated damages. To recover under this theory, the contractor must prove:

1. An excusable delay that an extension of time was requested according to the contract provisions.
2. The owner failed to grant an adequate or any extension of time.
3. That the owner made it clear that completion was required within the original contract period.
4. That adequate notice had been given by the contractor to the owner advising that he was treating the owner's actions as constructive acceleration.
5. And finally, the proof that there had been the actual insurance of additional costs (Carnell 2000; Rochester 2003).

If the contract provides for acceleration, payment should be based on the terms of the contract. If the contract makes no provision, the parties should agree the basis of payment before acceleration is commenced. In the absence of agreement, steps should be taken by party to have either the dispute or difference settled.

## References

Antil, J. M., & Woodhead, R. W. (1990). *Critical path methods in construction practice (Hardcover)*. New York: Wiley.

Carnell, N. J. (2000). *Causation and delay in construction disputes*. Oxford: Blackwell Science.

Driver, S. (1994). Provisions for acceleration in construction engineering contracts—the difficulties. *M.Sc. Thesis, Centre of Construction Law and Management*, King's College London.

FIDIC. (1999). *Fédération International des Ingénieurs—Conseils, conditions of contract (for building and engineering works designed by the employer) (The 'Red Book')*. Geneva, Switzerland: FIDIC.
Gerrard, R. (2007). Briefing: NEC contracts—Delivering 21st century construction. *Management Procurement and Law Proceedings of the Institution of Civil Engineers, 160*, 9–10.
Householder, J., & Rutland, H. (1990). Who owns floats? *Journal of Construction Engineering and Management, 116*(1), 130–133.
ICE 7th Edition. (1999, September). ICE conditions of contract, measurement version (7th ed.).
JCT 98 (1998). The JCT standard form of building contract, 1998 edition.
JCT 2005 (2005). The JCT standard form of building contract, 2005 edition.
Kallo, G. (1996). The reliability of critical path method (critical path method) techniques in the analysis and evaluation of delay claims. *Cost Engineering, 38*(5), 35–37.
Knoke, J. (1995). Avoiding delay claims with automated schedule reviews. In *Computing in Civil Engineering: Proceedings of the Second Congress* (pp. 1506–1513).
Latham, M. (1994). Final report of the government/industry review of procurement and contractual arrangements in the construction industry—constructing the team (1994, HMSO).
NEC3. (2005). *NEC engineering and construction contract* (3rd ed.). London: Thomas Telford Ltd. 2005.
Palles-Clark, R. A. (2003). The use of critical path analysis to prove claims for delay. *M.Sc. Thesis, School of Construction Law and Arbitration*, King's College, University of London.
Pickavance, K. (2005). *Delay and disruption in construction contracts* (3rd ed.), London.
PPC (2000). Project partnering contract.
Rochester, M. (2003). A practical approach to managing extension of time claims. *The Arbitrator and Mediator, 22*(2), 37.
Schumacher, L. (1996). An integrated and proactive approach for avoiding delay claims on major capital projects. *Cost Engineering, 38*(6), 37–39.
Seymour, N. (1995). Time-delay and causation. *M.Sc. Thesis, School of Construction Law and Arbitration*, King's College, University of London.
Thomas, P. (2004, February). Concurrent delay—finding equitable solutions, dissecting the doctrine of concurrent delay. Part 4 of 5, Pinnacle One E-Newsletter.
Wickwire, J., Warner, J., & Mark, R. (1999). *Construction scheduling, construction law handbook*, New York.

## Legal Court Cases

Amalgamated Building Contractors Ltd v Waltham Holy Cross UDC (1952) 2 All ER 452 CA.
Ascon Contracting Ltd v Alfred McAlpine Construction Isle of Man Ltd (1999) 66 Con LR 119; (2000) CILL 1583; (2000) 16 Const LJ 316, TCC.
Balfour Beatty v The Mayor and Burgess of the London Borough of Lambeth (2002) 12 April (TCC).
Balfour Building Ltd v Chestermount Properties Ltd (1993) 62 BLR 1 QBD.
FIDIC—International Federation of Consulting Engineers (The Global Voice of Consulting Engineers) http://www1.fidic.org/resources/contracts/farrow_march06.asp.
Fortec Constructors v United States 8 Cl. Ct. 490 (1985) Fortec.
Glenlion Construction Ltd v Guinness Trust (1987) 39 BLR 89 QBD.
Henry Boot Construction (UK) Ltd v Malmaison Hotel (Manchester) Ltd. (2000). CILL 1572 TCC. http://www.planningengineers.org/publications/legalcases.aspx.
John Barker Construction Ltd v London Portman Hotel Ltd. (1996). BLR 31.
John Driggs Company, Inc. (2005). COFC No 96-222C.
Joint Tribunal Central—The Standard Forms of Building Contract http://www.jctcontracts.com/JCT/contracts/98/9805sbc.jsp.

Motherwell Bridge Construction Ltd. (T/A Motherwell Storage Tanks) v (1) Micafil Vakkuumtechnik Ag (2) Micafil Ag (2002) 8 Con LR 44.
Multiplex Constructions (UK) Ltd v Honeywell Control Systems Ltd., TCC 6 March 2007.
Natkin & Co., George A. Fuller Co 347 F. Supp. 17 (W.D. Mo. 1972).
Peak Construction (Liverpool) Ltd v McKinney Foundations (1970) 1 BLR 114.
Preston-Brady Co. VABCA No 1892 (1991).
Royal Brompton Hospital NHS Trust v Hammond & Others (No 7) (2001) 76 Con LR 148.
R P Wallace v The United States COFC No 96-222C (2005).

# Chapter 4
# Mathematical Methods—Statistics and Forecasting

**Abstract** Given the improvement of data available, advanced modeling techniques, and computer programs, the design, construction, and the delivery of programs are likely to be more accessible with improved accuracy at every level of the program. This chapter will provide an introduction for statistics applications and forecasting methods to familiarize program managers with their applications as they have been proven applicable and apprehendable for various aspects of a program delivery. Statistics with its wide scope is a very relevant tool for program management. It is concerned with scientific methods for collecting, organizing, summarizing, presenting, and analyzing data as well as with drawing valid conclusions and making reasonable decisions on the basis of such analysis. In a narrower sense, the term statistics is used to denote the data themselves or numbers derived from the data. A forecast is a prediction, projection, or estimate of some future activity, event, or occurrence. When historical data are available, some proven statistical forecasting methods have been developed for using these data to forecast future demand. Such a method assumes that historical trends will continue, so management then needs to make any adjustments to reflect current changes in the construction industry and the management of programs.

## Statistics Analysis

Statistics appertaining to program management involves finding out the best sample and populations required to solve a wide range of problems. Thus, we speak of recruitment and employment statistics, on-site accident statistics and statistics related to types of construction, leverage risk provisions, and forecasting statistics analysis.

Programs are large and in many cases involve huge data. A program involving 400 schools in a developed country will involve mega studies related to the number of students per year, the age of the students, and the demographic distribution of the students which will involve the design, the size, and the type of schools. Then, it is



A.D. Haidar, *Construction Program Management – Decision Making and Optimization Techniques*, DOI 10.1007/978-3-319-20774-2_4

possible to ascertain the number of schools, their location, and the kind of schools required. For example, in a conservative Middle Eastern country, it is not possible to have mixed schools, while in certain African countries, the racial and religious demographic distribution is essential.

Statistics is concerned with using scientific methods in:

1. Collecting data relating to certain events or physical phenomena. Most data sets involve numbers
2. Organizing all collected data. Data sets are normally organized in either ascending order or descending order
3. Arranging data sets in logical and chronicle order for viewing and analyses
4. Summarizing the data to offer an overview of the situation
5. Developing a comprehensive way to present the data set
6. Analyzing the data set for the intended applications.

## Sampling Theory

In collecting data concerning the characteristics of a group of projects, sources of materials, available human resources, or design firms, and instead of examining the entire group, called the population, or universe, one examines a small part of the group, called a sample.

A population can be finite or infinite. The number of projects produced by one institution in one completion periodical plan is finite. The number of possible outcomes of successive tenders of an estimating department is infinite.

If a sample is representative of a population, important conclusions about the population can often be inferred from analysis of the sample. The phase of statistics dealing with conditions under which such inference is valid is called inductive statistics, or statistical inference. Because such inference cannot be absolutely certain, the language of probability is often used in stating conclusions. The phase of statistics that seeks only to describe and analyze a given group without drawing any conclusions or inferences about a larger group is called descriptive, or deductive, statistics.

Sampling theory is a study of relationships existing between a population and samples drawn from the population. It is useful in estimating unknown population quantities such as population mean and variance, often called population parameters, from acknowledge of corresponding sample quantities, such as sample mean and variance, often called sample statistics or briefly statistics. Sampling theory is also useful in determining whether the observed differences between two samples are due to chance variation or whether they are really significant. The answers involve the use of tests of significance and hypotheses that are important in the theory of decisions.

In order that the conclusions of sampling theory and statistical inference are valid, samples must be chosen so as to be representative of a population. A study of sampling methods of the related problems that arise is called the design of the experiment.

One way in which a representative sample may be obtained is by a process called random sampling according to which each member of a population has an equal chance of being included in the sample.

If we draw a number from an urn, we have the choice of replacing or not replacing the number into the urn before a second drawing. In the first case, the number can come up again and again where in the second it can only come up once. In the first instance, it is called sampling without replacement and the second sampling with replacement.

Populations are either finite or infinite. If for example, we draw 10 balls successively without replacement from an urn containing 100 balls, we are sampling from a finite population. A finite population in which sampling is with replacement can theoretically be considered infinite, since any number of numbers can be drawn without exhausting the population. For many practical purposes, sampling from a finite population that is very large can be considered to be sampling from an infinite population.

## Variables, Index, and Summation

A variable is symbol, such $X$, $Y$, $H$, or $B$, that can assume any of a prescribed set of values, called the domain of the variable. If the variable can assume only one variable, it is called a constant. A variable that can theoretically assume any value between two given values is called a continuous variable; otherwise, it is called a discrete variable.

The number $N$ of projects in a program, which can assume any of the values 0, 1, 2, 3, …, but cannot be 2.5 or 4.3 or 5.4 is a discrete variable.

Data that can be described by a discrete or continuous variable are called discrete data or continuous data, respectively.

In general, measurements give rise to continuous data, while enumerations, or counting, give rise to discrete data.

## Index

Let the symbol $X_j$ (read "$X$ sub $j$") denote any of the $N$ values $X_1, X_2, X_3, \ldots, X_N$ assumed by a variable $X$. The letter $j$ in $X_j$, which can stand for any of the numbers 1, 2, 3, …, N, is called a subscript, or index. Clearly, any letter other than $j$, such as $i$, $k$, $p$, $q$, or $s$, could have been used as well.

**Summation notice** ($\Sigma$ is the Greek capital letter sigma, denoting sum).

The symbol $\sum_{j=1}^{N} X_j$ is used to denote the sum of all $X_j$'s from $j = 1$ to $j = N$; by definition,

$$\sum_{j=1}^{N} X_j = X_1 + X_2 + X_3 + \cdots + X_N \tag{4.1}$$

We often denote this simply by $\Sigma X$ or $\Sigma X_j$.
For example,

$$\sum_{j=1}^{N} X_j Y_j = X_1 Y_1 + X_2 Y_2 + X_3 Y_3 + \cdots + X_N Y_N \tag{4.2}$$

$$\sum_{j=1}^{N} aX_j = aX_1 + aX_2 + aX_3 + \cdots + aX_N = a\sum_{j=1}^{N} X_j \tag{4.3}$$

## Averages or Measures of Central Tendency

An average is a value that is a typical representative of a set of data. Since such typical values tend to lie centrally within a set of data arranged according to magnitude, average is also called central tendency.

Several types of averages can be defined, the most common being the arithmetic mean, the median, and the mode. Each has advantages and disadvantages, depending on the data and the intended use.

The arithmetic mean, or the mean, of a set of $N$ numbers $X_1 + X_2 + X_3 + \cdots + X_N$ is denoted by $\overline{X}$ ($X$ bar) and is defined as

$$\overline{X} = \frac{X_1 + X_2 + X_3 + \cdots + X_N}{N} = \frac{\sum_{j=1}^{N} X_j}{N} = \frac{\sum X}{N} \tag{4.4}$$

The arithmetic mean of the numbers 8, 3, 5, 12, and 10 is

$$\overline{X} = \frac{8 + 3 + 5 + 12 + 10}{5} = \frac{38}{5} = 7.6 \tag{4.5}$$

If the numbers $X_1, X_2, X_3, \ldots, X_k$ occur $f_1, f_2, \ldots, f_k$ times, respectively (i.e., occur with frequencies $f_1, f_2, \ldots, f_k$), the arithmetic mean is

$$\overline{X} = \frac{f_1X_1 + f_2X_2 + f_3X_3 + \cdots + f_KX_K}{f_1 + f_2 + \cdots + f_K} = \frac{\sum_{j=1}^{K} f_jX_j}{\sum_{j=1}^{K} f_j} = \frac{\sum fX}{f} = \frac{\sum f_x}{N} \tag{4.6}$$

where $N = \Sigma f$ is the total frequency (i.e., the total number of cases).

For example, if 5, 8, 6, and 2 occur with frequencies 3, 2, 4, and 1, respectively, the arithmetic mean is

$$\overline{X} = \frac{(3)(5) + (2)(8) + (4)(6) + (1)(2)}{3 + 2 + 4 + 1} = \frac{15 + 16 + 24 + 2}{10} = 5.7 \tag{4.7}$$

Sometimes, we associate with the numbers $X_1$, $X_2$, $X_3$, …, $X_K$ certain weighing factors (or weights) $w_1$, $w_2$, $w_3$, …, $w_K$ depending on the significance or importance attached to the numbers. In this case

$$\overline{X} = \frac{w_1X_1 + w_2X_2 + w_3X_3 + \cdots + w_kX_K}{f_1 + f_2 + \cdots + f_K} = \frac{\sum wX}{w} \tag{4.8}$$

is called the weighed arithmetic mean.

For example, if an activity has a cost impact 3 times as much as other subactivities and this activity has a duration of 85 and subactivities durations of 70 and 90, the mean grade or average activity (cost–duration) is

$$\overline{X} = \frac{(1)(70) + (1)(90) + (3)(85)}{1 + 1 + 3} = \frac{415}{5} = 83 \tag{4.9}$$

## Median

The median of a set of numbers arranged in order of magnitude is either the middle value or the arithmetic mean of the two middle values.

For example:

The set of numbers 3, 4, 4, 5, 6, 8, 8, 8, and 10 has median 6.

The set of numbers 5, 5, 7, 9, 11, 12, 15, and 18 has median $0.5 \times (9 + 11) = 10$.

## Mode

The mode of a set of numbers is that value which occurs with the greatest frequency, which is the most common value. The mode may not exist, and even if it does exist, it may not be unique.

As for example:

The set 2, 2, 5, 7, 9, 9, 9, 10, 10, 11, 12, and 18 has mode 9.

The set 3, 5, 8, 10, 12, 15, and 16 has no mode.

The set 2, 3, 4, 4, 4, 5, 5, 7, 7, 7, and 9 has two modes, 4 and 7, and is called bimodal. A distribution that has only one mode is called unimodal.

## Dispersion or Variation

The degree to which numerical data tends to spread about an average value is called dispersion, or variation of the data. Various measures of this dispersion are available, the most common being the range, mean deviation, and standard deviation.

## Range

The range of a set of numbers is the difference between the largest and smallest numbers in the set.

For example:

The range of the set 2, 3, 3, 5, 5, 5, 8, 10, and 12 is 12 − 2 = 10. Sometimes, the range is given by simply quoting the smallest and largest numbers; in the above set, for instance, the range could be indicated as 2 to 12 or 2–12.

## Mean Deviation

The mean deviation, or average deviation, of a set of $N$ numbers $X_1, X_2, X_3, \ldots, X_K$ is abbreviated MD and is defined as

$$\text{Mean Deviation (MD)} = \frac{\sum_{j=1}^{N} \left|X_j - \overline{X}\right|}{N} = \frac{\sum \left|X - \overline{X}\right|}{N} = \left|\overline{X - \overline{X}}\right| \tag{4.10}$$

where $\overline{X}$ is the arithmetic mean of the numbers and $\left|X_j - \overline{X}\right|$ is the absolute value of the deviation of $X_j$ from $\overline{X}$. (The absolute value of a number is the number without the associated sign and is indicated by two vertical lines placed around the number).

## Standard Mean of Deviation Is

$$s = \sqrt{\frac{\sum_{j=1}^{N} \left(X_j - a\right)^2}{N}} \quad \text{where } a = \bar{X} \tag{4.11}$$

## Sampling Distribution

Consider all possible samples of size $N$ that can be drawn from a given population (either with or without replacement). For each example, we can compute statistics (such as the mean and the standard deviation) that will vary from sample to sample. In this manner, we obtain a distribution of the statistics that is called its sampling distribution.

The standard deviation of a sampling distribution of statistics is often called its standard error. When population parameters, such as standard deviation, are unknown, they may be estimated closely by their corresponding sample statistics, namely $s$, where

$$\widehat{S} = \sigma = s\sqrt{\frac{N}{N-1}} \quad \text{where } N \text{ is the sample size.} \tag{4.12}$$

## Unbiased Parameters

If the mean of the sampling distribution of as a statistics equals the corresponding population parameter, the statistics is called *an unbiased estimator of the parameter*; otherwise, it is called *a biased estimator*. The corresponding values of such statistics are called unbiased or biased estimates, respectively.

## Confidence Interval Estimates of Population Parameters

Let $\mu s$ and $\sigma s$ be the mean and the standard deviation (standard error), respectively, of the sampling distribution of statistics $S$. Then, if the sampling distribution of $S$ is approximately normal, we can expect to find an actual sample statistics $S$ lying in the intervals $\mu s - \sigma s$ to $\mu s + \sigma s$, $\mu s - 2\sigma s$ to $\mu s + 2\sigma s$, or $\mu s - 3\sigma s$ to $\mu s + 3\sigma s$ about 68.27, 95.45, and 99.73 % of the time, respectively.

**Table 4.1** Normal curve confidence levels

| Confidence level (%) | 99.73 | 99 | 98 | 96 | 95.45 | 95 | 90 | 80 | 68.27 | 50 |
|---|---|---|---|---|---|---|---|---|---|---|
| $z_c$ | 3.00 | 2.58 | 2.33 | 2.05 | 2.00 | 1.96 | 1.645 | 1.28 | 1.00 | 0.6745 |

At the same time, we can be confident of finding μs in the intervals $S - \sigma s$ to $S + \sigma s$, $S - 2\sigma s$ to $S + 2\sigma s$, or $S - 3\sigma s$ to $S + 3\sigma s$ about 68.27, 95.45, and 99.73 % of the time, respectively. Because of this, we call these respective intervals the, 68.27, 95.45, and 99.73 %, confidence intervals for estimating μs.

Similarly, $S \pm 1.96\sigma s$ and $S \pm 2.58\sigma s$ are the 95 and 99 % (or 0.95 and 0.99) confidence limits for $S$. The percentage confidence is often called the confidence level. The numbers 1.96 and 2.58, etc., in the confidence limits are called confidence coefficients or critical values and are donated by $z_c$. From confidence levels, we can find confidence coefficients and vice versa.

Table 4.1 shows the values of $z_c$ corresponding to various confidence levels used in practice. For confidence levels not presented in the table, the values of $z_c$ can be found from the normal curve area tables.

## Confidence Intervals of Means

If the statistics $S$ is the sample mean $\overline{X}$, then the 95 and 99 % confidence limits for estimating the population mean $\mu$ are given by $\overline{X} \pm 1.96\,\sigma_{\overline{X}}$ and $\overline{X} \pm 2.58\,\sigma_{\overline{X}}$, respectively.

More generally, the confidence limits are given by $\overline{X} \pm z_c\sigma_{\overline{X}}$
where $z_c$ can be read from the above table.

The confidence limits for the population mean are given by

$$\overline{X} \pm z_c \frac{\sigma}{\sqrt{N}} \tag{4.13}$$

If the sampling is either from an infinite population or with replacement from a finite population then the confidence limits for the population mean are given by

$$\overline{X} \pm z_c \frac{\sigma}{\sqrt{N}} \sqrt{\frac{N_p - N}{N_p - 1}} \tag{4.14}$$

For further reading, (see Hastie et al. 2001; Hill and Lewicki 2005; and Montgomery and Runger 2003).

# Forecasting

Forecasting is an important aid in effective and efficient planning. It can be defined as attempting to predict the future by using qualitative or quantitative means. It is greatly needed if lead time for decision time is big/several years which makes very helpful for programme management.

Some of the areas that forecasting currently plays an important role are:

1. Scheduling existing resources (Production, transportation, cash, personnel, equipment, materials...)
2. Acquiring additional resources
3. Determining what resources are required.

We always try to achieve the followings:

1. Application of a range of forecasting methods
2. Procedures for selecting the appropriate methods for a specific situation
3. Organizational support for applying and using formalized forecasting methods.

Types of forecasts are mainly:

**Quantitative**—These are techniques of varying levels of statistical complexity which are based on analyzing past data of the item to be forecast.

Time series (prediction of future is based on past values of a variable or past errors). It uses some form of mathematical or statistical analysis on past data arranged periodically, i.e., month, quarter, and year.
Causal methods/regression (assume that the factor to be forecasted exhibits a cause–effect relationship with one or more independent variables, i.e., sales is related to income, prices, advertising), refer to Fig. 4.1.

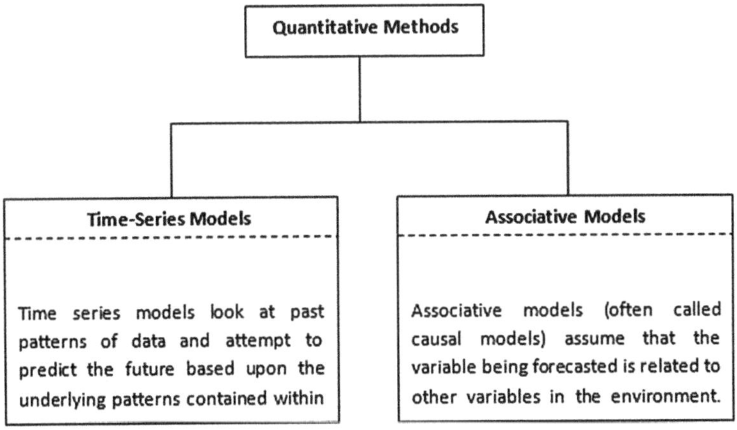


**Fig. 4.1** Quantitative methods

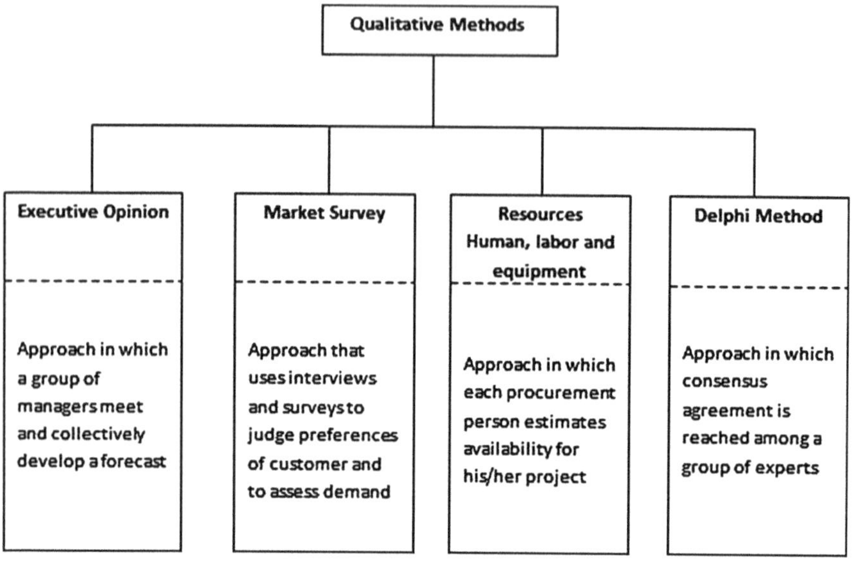


**Fig. 4.2** Qualitative methods

**Qualitative or technological methods**—These are techniques which are used when data are scarce.

> Exploration (begin with the past and present as their starting point and move towards the future in a heuristic manner, often looking at all available possibilities)Normative methods (starts with the future by determining future goals and objectives and then works backward to see whether these can be achieved, given the constraints, resources, and technologies available), refer to Fig. 4.2 (Makridakis et al. 1997).

## Delphi Method

This is a technique mainly used for longer term forecasting, designed to obtain expert consensus for a particular forecast. The procedure is that a panel of experts independently answer a sequence of questionnaires in which the responses to one questionnaire are used to produce the next questionnaire.

They are usually used for:

- Economic forecasts
  - Predict a variety of economic indicators, such as money supply, inflation rates, and interest rates.
- Technological forecasts
  - Predict the rates of technological progress and innovation.

- Demand forecasts
  - Predict the future demand for a company's products or services.

## Time Series Models

Time series analysis accounts for the fact that data points taken over time may have an internal structure (such as autocorrelation, trend, or seasonal variation) that should be accounted for, refer to Table 4.2.

## Least Square Estimates

For example, Table 4.3 summarizes the number of staff and theory of daily working hours.

Suppose that a project manager wants to know how much time he spends on site. The project manager might start by taking a sample of say 12 engineers and foremen. Factors to be considered are time of the day, day of the week, and sick days.

For convenience, error will be denoted by $e$, the data by $X$, and the pattern by $\overline{X}$. Error = data − pattern

$$e_i = X_i - \overline{X} \tag{4.15}$$

**Table 4.2** The different types of time series model

| Model | Description |
|---|---|
| Naïve | Uses last period's actual value as a forecast |
| Simple mean (average) | Uses an average of all past data as a forecast |
| Simple moving average | Uses an average of a specified number of the most recent observations, with each observation receiving the same emphasis (weight) |
| Weighted moving average | Uses an average of a specified number of the most recent observations, with each observation receiving a different emphasis (weight) |
| Exponential smoothing | A weighted average procedure with weights declining exponentially as data become older |
| Trend projection | Technique that uses the least squares method to fit a straight line to the data |
| Seasonal indexes | A mechanism for adjusting the forecast to accommodate any seasonal patterns inherent in the data |

**Table 4.3** Sample of number of staff and daily working hours

| Staff | Hour spent |
|---|---|
| 1 | 9 |
| 2 | 8 |
| 3 | 9 |
| 4 | 12 |
| 5 | 9 |
| 6 | 12 |
| 7 | 11 |
| 8 | 7 |
| 9 | 13 |
| 10 | 9 |
| 11 | 11 |
| 12 | 10 |

To examine the squared error

$$e_i^2 = (X_i - \overline{X})^2 \tag{4.16}$$

Summing these values (squared errors) for all 12 staff:

$$\phi = \sum_{i=1}^{12} e^2 = \sum_{i=1}^{12} (X_i - \overline{X})^2 \tag{4.17}$$

By taking the derivative of $\phi$, setting it equal to zero will minimize the sum of the squared errors.

$$\frac{\mathrm{d}\phi}{\mathrm{d}\overline{X}} = -2\sum_{i=1}^{12} (X_i - \overline{X}) \tag{4.18}$$

so that $\sum_{i=1}^{12} (X_i - \overline{X}) = 0$

$$\sum_{i=1}^{12} X_i - 12\overline{X} = 0 \quad \overline{X} = \frac{1}{12}\sum_{i=1}^{12} X_i \tag{4.19}$$

It is of course possible to minimize $\sum e_i$, $\sum e_i^3$, and $\sum e_i^4$, but minimizing MSE is the most popular as to avoid falling into the trap of having errors canceling each other if their sign differs or in the higher exponential, the computation can be difficult.

In Table 4.4, the following are calculated:

SSE = Sum of squared errors
MSE = Mean squared error

**Table 4.4** Sample of computation sum of squared errors and mean squared errors

| Estimated value | | 7 | | 9 | | 10 | | 12 | |
|---|---|---|---|---|---|---|---|---|---|
| Client | Amount spent | Error | Error squared | Error | Error squared | Error | Error squared | Error | Error squared |
| 1 | 9 | 2 | 4 | 0 | 0 | −1 | 1 | −3 | 9 |
| 2 | 8 | 1 | 1 | −1 | 1 | −2 | 4 | −4 | 16 |
| 3 | 9 | 2 | 4 | 0 | 0 | −1 | 1 | −3 | 9 |
| 4 | 12 | 5 | 25 | 3 | 9 | 2 | 4 | 0 | 0 |
| 5 | 9 | 2 | 4 | 0 | 0 | | | | |
| 6 | 12 | 5 | 25 | 3 | 9 | | | | |
| 7 | 11 | 4 | 16 | | | | | | |
| 8 | 7 | 0 | 0 | | | | | | |
| 9 | 13 | 6 | 36 | | | | | | |
| 10 | 9 | 2 | 4 | | | | | | |
| 11 | 11 | 4 | 16 | | | | | | |
| 12 | 10 | 3 | 9 | | | | | | |
| SSE | | | 144 | | 48 | | 36 | | 84 |
| MSE | | | 12 | | 4 | | 3 | | 7 |

Error = Amount spent − estimated value
MSE = SSE/12 which is the mean squared errors for estimates of staff attendance.

## Moving Average

Given a data set covering the last $N$ time periods:

$$X_1, X_2, X_3, \ldots X_{N-1} X_N \tag{4.20}$$

and a decision to use the first $T$ data points as the initialization set and the rest as a test.

$$\begin{array}{lll} X_1, X_2 \ldots & X_T & \text{(Initialisation Set)}\backslash \\ & X_N & \text{(Test Set)} \end{array} \tag{4.21}$$

The method of simple averages is to take the average of all data in the initialization set

$$\overline{X} = \sum_{i=1}^{T} X_i / T = F_{T+1} \tag{4.22}$$

**Table 4.5** The process of the moving average components is demonstrated in Table 4.5

| Time | Stored from last period | Input at this time | Output |
|---|---|---|---|
| $T$ | | $X_1, \ldots, X_T$ | $F_{T+1} = \sum_{i-1}^{T} X_i/T$ |
| $T+1$ | $T, F_T + 1$ | $X_T + 1$ | $F_{T+2} = \sum_{i-1}^{T+1} X_i/(T+1)$ |
| $T+2$ | $T+1, F_T + 1$ | $X_T + 1$ | $F_{T+3} = \frac{((T+1)xF_{T+2}+X_{T+2})}{(T+2)}$ |

as the forecast for period ($T + 1$). Then, as period's ($T + 1$) data become available, it is possible to compute an error:

$$e_{T+1} = X_{T+1} - F_{T+1} \tag{4.23}$$

For period ($T + 2$), the situation is

$$\begin{matrix} X_1, X \ldots X_{T+1} & \text{(Initialisation Set)} \\ X_{T+2} \ldots X_N & \text{(Test Set)} \end{matrix} \tag{4.24}$$

There is one more data point in the past history set, the new mean is

$$\overline{X} = \sum_{i=1}^{T+1} X_i/(T+1) = F_{T+2} \tag{4.25}$$

and the new error term, when $X_{T+2}$ becomes available, is (Table 4.5)

$$e_{T+2} = X_{T+2} - F_{T+2} \tag{4.26}$$

**Table 4.6** Sample of monthly forecast

| Month | Actual labor (number) | 3-month moving average | 6-month moving average | 12-month moving average |
|---|---|---|---|---|
| January | 450 | | | |
| February | 440 | | | |
| March | 460 | | | |
| April | 410 | 450 | | |
| May | 380 | 437 | | |
| June | 400 | 417 | | |
| July | 370 | 397 | 423 | |
| August | 360 | 383 | 410 | |
| September | 410 | 377 | 397 | |
| October | 450 | 380 | 388 | |
| November | 470 | 407 | 395 | |
| December | 490 | 443 | 410 | |
| January | 460 | 470 | 425 | 424 |

For the example shown in Table 4.6, lets assume for each of the year the number of labor available on a particular site;

Any month's forecast is the average of the proceeding $n$ months' labor. For example, the 3 monthly moving average forecasts were prepared as follows:

$$\begin{aligned}\text{April's forecast} &= \frac{\text{January count} + \text{February count} + \text{March count}}{3} \\ &= \frac{450 + 440 + 460}{3} = 450\end{aligned} \tag{4.27}$$

and so on.

## Exponential Smoothing

This is used to overcome the limitations of moving averages as equal weighting is given to each of the values in the calculations.

$$\begin{aligned}\text{New Forecast } &= \text{Old Forecast} \\ &\quad + \text{ a proportion of the forecast error}\end{aligned} \tag{4.28}$$

$$\begin{aligned}\text{New Forecast } &= \text{Old Forecast} \\ &\quad + \alpha(\text{Latest Observation } - \text{ Old Observation})\end{aligned} \tag{4.29}$$

where $\alpha$ is the smoothing constant. The value of $\alpha$ can be between 0 and 1. The higher the value of $\alpha$, the more sensitive the forecast becomes to current conditions. It will react more sensitively to current conditions.

The value of $\alpha$ can be between 0 and 1. An approximate equivalent of $\alpha$ values to the number of moving average is:

$$\begin{aligned}\text{March Forecast} &= \text{February forecast} + \alpha(\text{Feb count } - \text{ Feb forecast}) \\ &= 450 + 0.2(440 - 450) = 448\end{aligned} \tag{4.30}$$

By setting $\alpha$ at 0.2.

**Notes** Because no previous forecast was available, January count was used as February count; refer to Table 4.7.

**Table 4.7** Sample of exponential smoothing

| Month | Actual labor (number) | $\alpha = 0.2/3$ months MA | $\alpha = 0.8/3$ months MA |
|---|---|---|---|
| January | 450 | | |
| February | 440 | 450 | 450 |
| March | 460 | 448 | 442 |
| April | 410 | 450.4 | 456.4 |
| May | 380 | 442.32 | 419.28 |
| June | 400 | 429.86 | 387.86 |
| July | 370 | 423.89 | 397.57 |
| August | 360 | 413.11 | 375.51 |
| September | 410 | 402.49 | 363.102 |
| October | 450 | 403.99 | 400.62 |
| November | 470 | 413.19 | 440.12 |
| December | 490 | 424.55 | 464.02 |
| January | 460 | 437.64 | 484.80 |

## Time Series Analysis—Decomposition

Observation taken over time often contains the four following characteristics:

1. A long-term trend (*T*). It is the long-term tendency of the whole series to rise or fall
2. Seasonal variation (*S*). It is the short periodic fluctuations in values due to different circumstances
3. Cyclic variations (*C*). These are medium-term changes caused by factors which apply for a while, then go away, and then come back in repetitive cycle
4. Random or residual variations (*R*). These are nonrecurring random variations.

In order to make reasonably accurate forecasts, it is often necessary to separate out the above characteristics from the raw data. This is known as time series decomposition or time series analysis.

## Additive Model

Time series value = $T + S + C + R$ where *S*, *C*, and *R* are expressed as absolute values.

## Multiplicative Model

Time series value = $(T * S * C * R)$, where *S*, *C*, and *R* are expressed as percentages or proportions.

It is helpful to discuss and to classify the smoothing methods discussed. The group called "averaging methods" conforms to the conventional understanding of what an average is—equally weighed observations.

The next method is the moving average of the latest $n$ observations and we end up with unequally weighed averages and can be used within a forecasting method sometimes known as "linear" moving averages.

The other group of methods applies an unequal sets of weights to past data, and because the weights typically decay in an exponential manner from the most recent to the most distant data point, the methods are known as exponential smoothing methods. All methods in this group require that certain parameters be defined, and these parameter values lie between 0 and 1.

Analogous to the single moving average is the single exponential smoothing (SES) method, for which just one parameter needs to be defined. Analogous to double moving averages, there are two double exponential smoothing equations—involving two exponential smoothing equations.

Holt's method makes use of two different parameters for the two separate exponential smoothing involved.

## Double Moving Averages

This is a variation on the moving average procedure that is intended to deal better with trends. To mitigate against a systematic error that occurs if moving averages are applied to data with trend, the method of linear moving averages has been developed. The basis of this method is to calculate a second moving average. The double moving average is a moving average, and in symbols, it would be denoted by MA(M × N) where we mean an M-period MA of an $N$-period MA. Shown below is an example of MA (3 × 3)—a 3-period moving average.

The linear moving average forecasting procedure thus involves 3 aspects:

- The use of single moving average at time $t$ denoted by $S'_t$
- An adjustment, which is the difference between the single and the double moving average at time $t$ (denoted $S'_{t.} - S''_{t.}$), and
- An adjustment for trend from period $t$ to period $t + 1$ (or to period $t = m$ if we want to forecast $m$ periods ahead). ( Refer to Table 4.8).

## Forecasting a Series with Trend Using a Linear Moving Average

Note the following:

1. The first 3-period moving average is known as soon as the third data point ($X_3$) is known

**Table 4.8** Sample of double moving average

| 1 | 2 | 3 | 4 | 5 | 6 | 7 | 8 |
|---|---|---|---|---|---|---|---|
| Period | Actual value | Single moving average $N = 3$ | Error difference (2) − (3) | Double moving average $N = 3$ | Error difference (3) − (5) | Forecast (3) + (6) + Trend | Error difference (2) − (7) |
| 1 | 2 | | | | | | |
| 2 | 4 | | | | | | |
| 3 | 6 | 4 | 2 | | | | |
| 4 | 8 | 6 | 2 | | | | |
| 5 | 10 | 8 | 2 | 6 | 2 | | |
| 6 | 12 | 10 | 2 | 8 | 2 | 12 | 0 |
| 7 | 14 | 12 | 2 | 10 | 2 | 14 | 0 |
| 8 | 16 | 14 | 2 | 12 | 2 | 16 | 0 |
| 9 | 18 | 16 | 2 | 16 | 2 | 18 | 0 |
| 10 | 20 | 18 | 2 | 16 | 2 | 20 | 0 |
| 11 | | | | | | 22 | |

2. We place the first MA in column 3 opposite to time period 3
3. Similarly, as soon as the fifth data ($X_5$) are known, there will be three MA(3) values available (4, 6, and 8) and so the first MA(3 × 3) can be computed in column 5 and placed against period 5
4. The differences in column 4 and column 6 are identical (for this special data set), and so by the judicious use of the single moving average (column 3) and the difference between single and double moving averages (column 6), a forecast for the next period can be derived.

$$\begin{aligned} \text{F6} &= (\text{MA}(3) \text{ at period } 5) + ((\text{MA}(3)) - \text{MA}(3 \times 3) \text{ at period } 5 \\ &\quad + (\text{trend from period 5 to period 6}) \\ &= (8) + (2) + (2) = 12 \end{aligned} \tag{4.31}$$

The discussion may be generalized as follows: The general linear moving average procedure may be described by the following equations:

$$S'_t = X_t + X_{t-1} + X_{t-2} + \cdots + X_{t-N+1}/N. \tag{4.32}$$

This equation assumes that we are standing at time period $t$ and looking over the last $N$ known values. The single MA($N$) is denoted by $S'_t$.

$$S''_t = S'_t + S'_{t-1} + S'_{t-2} + \cdots + S'_{t-N+1}/N \tag{4.33}$$

This equation assumes that all the single moving averages ($S'$) have been computed and we compute the $N$-period moving average of the $S'$ values. The double moving averages are denoted as $S''$.

$$a_t = S'_t + (S'_t - S''_t) = 2S'_t - S''_t \tag{4.34}$$

This equation refers to the adjustment of the single MA, $S'_t$, by the differences ($S'_t - S''_t$).

$$b_t = 2(S'_t - S''_t)/(N - 1) \tag{4.35}$$

defines the estimate of trend from one period to the next.

$$F_{t+m} = a_t + b_t m \tag{4.36}$$

This equation shows how to obtain forecasts for $m$ periods ahead of $t$. The forecast for $m$ periods ahead is $a_t$—which is the adjusted smoothed value for period $t$—plus $m$ times the trend component $b_t$.

## Double Exponential Smoothing: Holt's Two-Parameter Method

Holt's linear exponential smoothing provides greater flexibility, since it allows the trend to be smoothed with different parameter that used on the original series. The forecast for Holt's linear exponential smoothing is found using two smoothing constants (with values between 0 and 1) and three equations: (Refer to Table 4.9)

$$S_t = \alpha X_t + (1 - \alpha)(S_{t-1} + b_{t-1}) \tag{4.37}$$

$$b_t = \gamma(S_t - S_{t-1}) + (1 - \gamma)b_{t-1} \tag{4.38}$$

$$F_{t+m} = S_t + b_t m \tag{4.39}$$

## Curve Fitting

To determine an equation that connects variables, a first step is to collect data that show corresponding values of the variables under consideration.

**Table 4.9** Sample of linear moving average

| | 1 | 2 | 3 | 4 | 5 | 6 |
|---|---|---|---|---|---|---|
| Period | Inventory balance product E12 | 4-month moving average of (1) | 4-month moving average of (2) | Value of $a$ | Value of $b$ | Value of $a + b\ (m)$ when $m = 1$ |
| 1 | 140.00 | | | | | |
| 2 | 159.00 | | | | | |
| 3 | 136.00 | | | | | |
| 4 | 157.00 | 148.00 | | | | |
| 5 | 173.00 | 156.25 | | | | |
| 6 | 131.00 | 149.25 | | | | |
| 7 | 177.00 | 159.5 | 153.25 | 165.75 | 4.166 | |
| 8 | 188.00 | 167.25 | 158.06 | 176.43 | 6.125 | 169.91 |
| 9 | 154.00 | 162.5 | 159.62 | 165.37 | 1.916 | 182.56 |
| 10 | 179.00 | 174.5 | 165.93 | 183.06 | 5.708 | 167.29 |
| 11 | 180.00 | 175.25 | 169.87 | 180.62 | 3.583 | 188.77 |
| 12 | 160.00 | 168.25 | 170.12 | 166.37 | −1.250 | 184.20 |
| 13 | 182.00 | 175.25 | 173.31 | 177.18 | 1.291 | 165.12 |
| 14 | 192.00 | | | | | |
| 15 | 224.00 | | | | | |
| 16 | 188.00 | | | | | |
| 17 | 198.00 | | | | | |
| 18 | 206.00 | | | | | |
| 19 | 203.00 | | | | | |
| 20 | 238.00 | | | | | |
| 21 | 228.00 | | | | | |
| 22 | 231.00 | | | | | |
| 23 | 221.00 | | | | | |
| 24 | 259.00 | | | | | |
| 25 | 273.00 | | | | | |
| 26 | | | | | | 266.31 |

Plotting $(X_1, Y_1), (X_2, Y_2), \ldots, (X_N, Y_N)$ on a rectangular coordinate system results in a diagram called a scatter diagram.

From the scatter diagram, it is often possible to visualize a smooth curve that approximates the data. Such a curve is called an approximating curve. If the data appear to be approximated well by a straight line, we have a linear relationship.

The general problem of finding equations of approximating curves that fit given sets of data is called curve fitting.

The equation of a straight line

$$Y = a_o + a_1 X \tag{4.40}$$

Given any 2 points ($X_1$, $Y_1$) and ($X_2$, $Y_2$) on the line, the constants $\underline{a}_0$ and $a_1$ can be determined. The resulting equation of the line can be written

$$Y - Y_1 = \left(\frac{Y_2 - Y_1}{X_2 - X_1}\right)(X - X_1) \text{ or } Y - Y_1 = m(X - X_1) \tag{4.41}$$

where

$$m = \frac{Y_2 - Y_1}{X_2 - X_1} \tag{4.42}$$

is called the slope of the line and represents the change in $Y$ divided by the corresponding change in $X$.

So in $Y = a_0 + a_1 X$

$a_1$ is the slope $m$, and $a_0$ is the value of $Y$ when $X = 0$, where $Y$ is called the $Y$ intercept.

## Method of Least Squares

Of all curves approximating a given set of data points, the curve having the property that $D_1^2 + D_2^2 + \cdots + D_2^n$ is a minimum is called a best fitting curve.

The least square line of the set of points ($X_1$, $Y_1$), ($X_2$, $Y_2$), …, ($X_N$, $Y_N$) has the equation

$$Y = a_0 + a_1 X \tag{4.43}$$

where the constants $a_0$ and $a_1$ are determined by solving the equations simultaneously

$$\sum Y = a_0 N + a_1 \sum X \tag{4.44}$$

$$\sum XY = a_0 \sum X + a_1 \sum X^2 \tag{4.45}$$

which are called the normal equations for the least squares line. The constants $a_0$ and $a_1$ of the above equations can be found as

$$a_0 = \frac{(\sum Y)(\sum X^2) - (\sum X)(\sum XY)}{N \sum X^2 - (\sum X^2)} \tag{4.46}$$

$$a_1 = \frac{N\sum XY - (\sum X)(\sum Y)}{N\sum X^2 - (\sum X^2)} \tag{4.47}$$

## Regression

Often, on the basis of sample data, we wish to estimate the value of variable $Y$ corresponding to a given value of a variable $X$. This can be accomplished by estimating the value of $Y$ from a least squares curve that fits the sample data. The resulting curve is called a regression curve of $Y$ on $X$, since $Y$ is estimated from $X$. (Refer to Table 4.10)

To obtain the equation of line, choose any two line, such as $P$ and $Q$; the coordinates of $P$ and $Q$ are (0, 1) and (12, 7.5) (Table 4.11).

$$\begin{aligned} 1 &= a_0 + a_1(0) \\ 7.5 &= a_0 + 12a_1 \\ a_0 &= 1 \\ a_1 &= 6.5/12 = 0.542 \end{aligned} \tag{4.48}$$

**Table 4.10** Sample of regression

| $X$ | $Y$ |
|---|---|
| 1 | 1 |
| 3 | 2 |
| 4 | 4 |
| 6 | 4 |
| 8 | 5 |
| 9 | 7 |
| 11 | 8 |
| 14 | 9 |

**Table 4.11** Sample of regression computation

| $X$ | $Y$ | $X^2$ | $XY$ | $Y^2$ |
|---|---|---|---|---|
| 1 | 1 | | | |
| 3 | 2 | | | |
| 4 | 4 | | | |
| 6 | 4 | | | |
| 8 | 5 | | | |
| 9 | 7 | | | |
| 11 | 8 | | | |
| 14 | 9 | | | |
| $\sum X =$ | $\sum Y =$ | $\sum X^2 =$ | $\sum XY =$ | $\sum Y^2 =$ |

**Table 4.12** Sample of regression computation

| | $X$ (quarters) | $Y$ | $XY$ | $X^2$ |
|---|---|---|---|---|
| Year 1 | 1 | 20 | 20 | 1 |
| | 2 | 32 | 64 | 4 |
| | 3 | 62 | 186 | 9 |
| | 4 | 29 | 116 | 16 |
| Year 2 | 5 | 21 | 105 | 25 |
| | 6 | 42 | 252 | 36 |
| | 7 | 75 | 525 | 49 |
| | 8 | 31 | 248 | 64 |
| Year 3 | 9 | 23 | 207 | 81 |
| | 10 | 39 | 390 | 100 |
| | 11 | 77 | 847 | 121 |
| | 12 | 48 | 576 | 144 |
| Year 4 | 13 | 27 | 351 | 169 |
| | 14 | 39 | 546 | 196 |
| | 15 | 92 | 1380 | 225 |
| | 16 | 53 | 848 | 256 |
| | $\sum X = 136$ | $\sum Y = 170$ | $\sum XY = 6661$ | $\sum X^2 = 1496$ |

Thus, the required equation is:

$$Y = 1 + 0.542 \tag{4.49}$$

and using Eq. 4.49 we get Table 4.12 inserted data.
Least square equations are:

$$\sum y = an + b \sum x \tag{4.50}$$

$$\sum xy = a \sum x + b \sum x^2$$
$$710 = 16a + 136b$$
$$626 = 340b$$
$$b = 1.84$$
$$a = 28.74$$

$$\mathbf{Trendline = y = 28.74 + 1.84x} \tag{4.51}$$

Use the trend line to calculate the estimate (being production, activity time, labor number, etc.) for each quarter. The estimate for the first quarter for the first year is: (Refer to Table 4.13)

**Table 4.13** Sample of regression computation

| | *X* (quarters) | *Y* | Estimated sales | Actual percentage estimated |
|---|---|---|---|---|
| Year 1 | 1 | 20 | 30.58 | 65 |
| | 2 | 32 | 32.42 | 99 |
| | 3 | 62 | 34.26 | 181 |
| | 4 | 29 | 36.10 | 80 |
| Year 2 | 5 | 21 | 37.94 | 55 |
| | 6 | 42 | 39.78 | 106 |
| | 7 | 75 | 41.62 | 180 |
| | 8 | 31 | 43.46 | 71 |
| Year 3 | 9 | 23 | 45.30 | 51 |
| | 10 | 39 | 47.14 | 83 |
| | 11 | 77 | 48.98 | 157 |
| | 12 | 48 | 50.82 | 94 |
| Year 4 | 13 | 27 | 52.66 | 51 |
| | 14 | 39 | 54.50 | 72 |
| | 15 | 92 | 56.34 | 163 |
| | 16 | 53 | 58.18 | 91 |

**Table 4.14** Average percentage variations

| Quarter 1 (%) | Quarter 2 (%) | Quarter 3 (%) | Quarter 4 (%) |
|---|---|---|---|
| 65 | 99 | 181 | 80 |
| 55 | 106 | 180 | 71 |
| 51 | 83 | 157 | 94 |
| 51 | 72 | 163 | 91 |
| 222 | 360 | 681 | 336 |
| 56 % | 90 % | 170 % | 84 % |

$$\text{Estimate} = 28.74 + 1.84(1) = 30.58 \tag{4.52}$$

$$\frac{\text{The actual value of } X\,\%}{\text{Estimates}} = \frac{20}{30.58} = 65\,\% \tag{4.53}$$

Compute the percentage variations to find the average seasonal variations. (Refer to Table 4.14)

On average, the first quarter of each year will be 56 % of the value of the trend. And the sum of averages divided by 4 should be equal to 100 %.

Prepare the final forecasts based on the above trend lines. (Refer to Table 4.15)

The forecasts for the first and second quarters are:

$$\text{Forecast } Q_1 = 30.58 \times 56\,\% = 17.112 \tag{4.54}$$

**Table 4.15** Sample of seasonal adjusted forecast

| | $X$ (quarters) | $Y$ (sales) | Seasonally adjusted forecast |
|---|---|---|---|
| Year 1 | 1 | 20 | 17.12 |
| | 2 | 32 | 29.18 |
| | 3 | 62 | 58.24 |
| | 4 | 29 | 30.32 |
| Year 2 | 5 | 21 | 21.24 |
| | 6 | 42 | 35.80 |
| | 7 | 75 | 70.75 |
| | 8 | 31 | 36.51 |
| Year 3 | 9 | 23 | 25.37 |
| | 10 | 39 | 42.43 |
| | 11 | 77 | 83.27 |
| | 12 | 48 | 42.69 |
| Year 4 | 13 | 27 | 29.49 |
| | 14 | 39 | 49.05 |
| | 15 | 92 | 95.78 |
| | 16 | 53 | 48.87 |

$$\text{Forecast } Q_2 = 32.42 \times 90\,\% = 29.18 \tag{4.55}$$

To forecast new trend, we can use the regression equation. For quarter (17)

$$\begin{aligned}\text{Basic Trend} &= 28.74 + 1.84(17)\\ &= 60.02\end{aligned} \tag{4.56}$$

Seasonal adjustment for first quarter = 56 % (Table 4.15)

$$\text{Adjusted forecast} = 60.02 \times 56\,\% = 33.61 \tag{4.57}$$

For further reading, see Makridakis et al. (1997).

## Forecasting Methods' Uses and Disadvantages

Surveys provide information about which forecasting methods are used by and the criteria used to evaluate them. However, it is also important to consider how the methods and criteria impact decisions.

1. Decision makers often ignore forecasts that are surprising or unpleasant
2. Forecasts were ignored if they conflicted with management's prior beliefs
3. Decision makers may be subject to various biases that lead them to revise the forecasts

4. The advocacy and illusion of control biases in new methods and products forecasting are given
5. Managers are reluctant to accept forecasts because they do not understand the methods used to obtain them
6. Users of forecasts in their organization seldom understood the forecasting methods that were used.

There are many factors to consider when forecasting for marketing decision making. For instance, one could consider:

1. What forecasts are needed (e.g., inflation, resource availability, materials required, cash flow...)
2. What situation exists (e.g., stage of the program life cycle, state of the construction industry, degree of regulation in the industry...)
3. What forecast horizon is appropriate (e.g., current status, short, medium, or long range)
4. What data are relevant and available
5. With what frequency must the forecast be prepared.
6. Who will prepare the forecast, and how much time and resources will be committed to the task
7. Who will use the forecast and in what manner
8. What process is to be used (e.g., how are the data to be gathered and analyzed, and how is the forecast to be presented)
9. When is the forecast needed
10. What uncertainty measures are needed.

## Applications of Statistics and Forecasting Methods for Program Management

### *Forecasting of Tenders for the Construction of Programs*

Statistics and forecasting techniques are a useful tool to establish an accurate-based model for the forecasting of the tendered price for the construction process for individual projects and the program in general.

Contractors tend to place low bids when tenders are invited for domestic public construction projects and programs. Overcompetition can lead to price wars to win a tender, which can in turn seriously affect the quality of construction. Forecasting techniques will assist the client being the public sector or private sector to determine what would be a reasonable reserve price or award. Multiple factors in the regression for the tender price prediction include the contract schedule, the budget price, and the tender bond. This will eliminate the risks of:

- Bid winners are forced to adopt the business pattern of subcontracting in order to split their risks, and transfer costs and responsibilities. The subcontracting pattern may have several negative consequences, such as the degradation on the quality of construction, and the difficulty in managing multiple projects at the same time
- For the client and stakeholders, it is a very important matter to select a contractor (s) in excellent financial condition and with the right management, reliability for planning, organizing, control, and human resource management skills to construct a program
- In addition, in order to solve the problem of overcompetitiveness in the construction, industry clients needed to be informed in advance of their likely future financial commitments and cost implications with the design evolution. This requires the estimation of building costs which is done based on historical cost data updated by the forecast tender price index.

Therefore, during tender, statistics and forecasting methods are essential to:

1. Study the real value of the investment to micro-level budgeting such as the forecast price of the construction program
2. Acquire accurate price indexes as tendering is an important task for construction companies
3. Analyze the tendering results which will have a great influence on the operating performance and profits of the construction company
4. Estimate money, time, and manpower which must be invested to submit tenders
5. Evaluate tenders properly so contractors are not carrying too much of a risk.

Therefore, it is very important for construction companies to offer suitable prices for tenders for construction projects they are about to bid on based on the price of awards from previous tenders (Ng et al. 2001; Wong et al. 2004; Wong and Ng 2010).

## Forecast Models for Actual Construction Time and Cost

Forecasting models have been developed for the actual construction time and cost when client sector, contractor selection method, contractual arrangement, project type are known, and while contract period and contract sum are estimated. Since these models for time and cost are dependent on the contract period and contract sum being known, it is necessary to investigate the effects in situations where these have to be estimated. The effects of different project type, contractor selection method, and contractual arrangement are also important.

The actual construction time and cost of a construction program may be affected by the client, planning, and contractual characteristics, and in many cases can be very different from the contract time and cost. Forecasting analysis is used for the development of the model for actual construction time forecast for different aspects of the program construction mainly:

1. Client sector
2. Contractor selection method
3. Contractual arrangement
4. Program type
5. Contract period and contract sum.

Statistical analysis can achieve the best model for actual construction time prediction to comprise:

1. The independent variable log contract time
2. Lump sum procurement
3. Nonstandard contractor selection (Odeyinka et al 2002; Skitmore and Ng 2003).

## *Forecasting Models to Predict Manpower Demand*

Analysis of manpower supply and demand has a long history and served as an important tool in the area of human resources planning. Manpower is regarded as a crucial resource element upon which the construction industry depends. Shortages in any particular category can result in disruptions in output and reduce productivity, whereas surplus of skilled workers can cause losses in the program overall cost. Manpower forecasting is, therefore, needed to facilitate the construction of programs and to prevent losses caused by attempts to undertake construction when and where the resources are not available. Forecasting manpower requirements has been useful for economic planners, policy makers, and program managers in order to avoid the imbalance of skills in the labor market. Forecasting precedes other methodologies by its dynamic nature and sensitivity to a variety of factors affecting the level and structure of employment.

Given the improvement of the data available, advanced modeling techniques, and computer programs, manpower planning is likely to be more accessible with improved accuracy at every level of the program.

The main purpose of manpower forecasting is as follows:

1. The latest employment and manpower demand estimating methods by examining their rationale, strength, and constraints
2. It aims to identify enhancements for further development of manpower forecasting model for the construction industry and compare the reliability and capacity of different forecasting methodologies
3. Given the improvement of the data available, advanced modeling techniques, and computer programs, manpower planning is likely to be more accessible with improved accuracy at every level of the society (Wong et al. 2004; Bartholomew et al. 1991; Agapiou et al. 1995; Willems 1996).

## *A Decision Support Model for Construction Program Management—An Overview*

The excessive level of program failures and their association with financial difficulties has placed financial management in the forefront of the construction of program imperatives. This has highlighted the importance of cash flow forecasting and management which has given rise to the development of several forecasting models. The traditional approach to the use of program financial models has been largely project-oriented perspective. However, the dominating role of "project economics" in shaping "program economics" tends to place the program strategy at the mercy of the projects.

The use of forecasting models should be regarded as tools for driving the program in the direction of organization aim. The separation of data from the mathematical expression enables unlimited expansion of the application of the model to all possible scenarios without the need to restructure the mathematical expression. The descriptive data that define each program can be summarized as follows:

1. Program characteristics
   (a) Finance
   (b) Design
   (c) Contractor(s) selection
   (d) Contract documents
   (e) Execution project 1 to project *n*
   (f) Completion and handover
   (g) Occupancy
   (h) Facility management.
2. Human resources
   (a) Recruiting foremen and labors for different skills
   (b) Mobilizing and distributing the manpower to the different sites
   (c) Organize insurances such as
      (i) Contractor plant and machinery insurance
      (ii) Comprehensive automobile insurance
      (iii) All risk cargo insurance
      (iv) Workmen compensation insurance
      (v) Health insurance.
3. Supervision and Engineering
   (a) To prepare construction cost estimates based on master plan and bills of quantities (if applicable) and establish overall construction cost budget
   (b) Advise on the cost planning and budget management of the programt
   (c) Prepare detailed estimates upon availability of schematic design to be refined progressively with the development of design

(d) Check the quotation from subcontractors and suppliers submitted to the contractor for variations and add new rates that are not in the priced bills of quantities upon the request of the employer and contractor(s) joint tendering work team
(e) Participate in value engineering exercise initiated by the contractor or the employer to verify its cost-effectiveness
(f) Review the prices from specialist consultants whose services are commissioned by the contractor and that are intended to be part of the total construction cost
(g) Verify the prepared bills of quantities and other tender document.

4. Technical support

   (a) Design shop drawings
   (b) Quantity surveying
   (c) Invoicing
   (d) Planning
   (e) Quality assurance
   (f) Quality control
   (g) Safety
   (h) Contract management
   (i) Procurement.

For any program, defined broadly in terms of the above characteristics, the forecasting models will predict the value of the profile variables. The influence of each characteristic varies for each case which is determined by the significance level produced by the aforementioned regression models. Once a forecast is generated, the analysts can use their general experience to improve the forecast. Also, they can use their knowledge about the program to further refine the forecast (Khosrowshahi and Kaka 2007; Odeyinkal and Lowe 2000).

## References


Agapiou, A., Price, A. D. F., & McCaffer, R. (1995). Planning future construction skill requirements: Understanding labour resource issues. *Construction Management and Economics, 13*(2), 149–161. 1995.

Bartholomew, D. J., Forbes, A. F., & McClean, S. I. (1991). *Statistical techniques for manpower planning* (Wiley series in probability and statistics—Applied probability and statistics section) Hardcover.

Hastie, T., Tibshirani, R., & Friedman, J. H. (2001). *The elements of statistical learning, data mining, inference and prediction*. New York: Springer Series in Statistics.

Hill, T., & Lewicki, P. (2005). *Statistics: Methods and applications paperback*.

Khosrowshahi, F., & Kaka, A. (2007). A decision support model for construction cash flow management. *Computer-Aided Civil and Infrastructure Engineering, 22*(2007), 527–539.

Makridakis, S. R., Wheelwright, S. C., & Hyndman, R. J. (1997). *Forecasting: Methods and applications hardcover*.

Montgomery, D. C., & Runger, G. C. (2003). *Applied statistics and probability for engineers* (3rd ed.). New York: Wiley.

Ng, S. T., Fan, R. Y. C., & Wong, J. M. W. (2001). An econometric model for forecasting private construction investment in Hong Kong. *Construction Management and Economics, 29*(5), 519–534.

Odeyinka, H. A., Lowe, J. G., & Kaka, A. (2002). A construction cost flow risk assessment model. In: D. Greenwood (Ed.), *Paper Published in the Proceedings of the 18th Annual ARCOM Conference held at Northumbria University*, Newcastle (pp. 3–12).

Odeyinkal, H. A., & Lowe, J. G. (2000). An assessment of risk factors involved in modelling cash flow forecast. *Paper Published in the Proceedings of the 16th Annual Association of Researchers in Construction Management (ARCOM) Conference Held at Glasgow Caledonian University* (pp. 557–565).

Skitmore, M., & Ng, S. T. (2003). Forecast models for actual construction time and cost. *Building and Environment, 38*(8), 1075–1083.

Willems, E. (1996). *Manpower forecasting and modeling replacement demand: An overview.* ROA-W-1996/4E, Maastricht.

Wong, J., Chan, A., Chiang, Y. H. (2004). A critical review of forecasting models to predict manpower demand. *Construction Economics and Building*, *4*(2).

Wong, J., & Ng, S. T. (2010). Forecasting construction tender price index in Hong Kong using vector error correction model. *Construction Management and Economics, 28*(12), 1255–1268. doi:10.1080/01446193.2010.487536.

# Chapter 5
# Operations Research and Optimization Techniques

**Abstract** This chapter will look at the principles of operations research and quantitative methods that are most accessible and suitable for program managers. Operations research is, in principle, the application of scientific methods, techniques, and tools for solving problems involving the operations of a system in order to provide those in control of the system with optimum solutions to problems. Put simply, it is a systematic and analytical approach to decision making and problem solving. This chapter provides an overview of operations research, its approach to solving problems, and some examples of successful applications. From the standpoint of a program manager, operations research is a tool that can do a great deal to improve productivity, assist in decision making, and optimize solutions. Therefore, the potential rewards can be enormous. Optimization techniques are also explained in this chapter to help program managers understand their importance. The last part of the chapter will look at linear programming methods and applications for construction, as this is the most widely applicable field for these types of problems. Linear programming can be used to allocate, assign, schedule, select, or evaluate whatever possibilities limited resources possess for different jobs. It has been used extensively in construction-related problems, where it can deduce the most profitable methods of allocating resources.

## Operations Research—Introduction

A common misconception, held by many, is that operations research is a collection of mathematical tools. While it is true that it uses a variety of mathematical techniques, operations research has a much broader scope. Operations research does not preclude the use of human judgment or non-quantifiable reasoning; rather, the latter are viewed as being complementary to the analytical approach. One should, thus, view operations research not as an absolute decision-making process, but as an aid to making good decisions (Tam et al. 2007).



A.D. Haidar, *Construction Program Management – Decision Making and Optimization Techniques*, DOI 10.1007/978-3-319-20774-2_5

Operations research plays an advisory role by presenting the program manager or a decision maker with a set of sound, scientifically derived alternatives. However, the final decision is always left to the human being, who has knowledge that cannot be exactly quantified, and who can temper the results of the analysis to arrive at a sensible decision.

Operations research is applied to problems that concern how to conduct and coordinate the operations (i.e., the activities) within an organization. It has been applied extensively in diverse areas such as manufacturing, transport, construction, telecommunications, planning, military operations, and management systems. It is a vital tool for program managers, and without its application, construction programs are at risk of being less effectively developed and managed.

In particular, the process begins by carefully observing and formulating the problem, including gathering all relevant data. The next step is to construct a scientific (typically mathematical) model that attempts to abstract the essence of the real problem. It is, then, hypothesized that this model is a sufficiently precise representation of the essential features of the situation and that the solutions obtained from the model are also valid.

Next, suitable experiments are conducted to test this hypothesis, modify it as needed, and eventually verify some form of the hypothesis. Thus, in a certain sense, operations research involves creative scientific research into the fundamental properties of operations. An additional characteristic is that operations research frequently attempts to find a best solution (referred to as an optimal solution) for the problem under consideration. Rather than simply improving the status quo, the goal is to identify a best possible course of action.

Defining operations research is a difficult task as its boundaries and content are not yet fixed. It can be regarded as the use of mathematical and quantitative techniques to substantiate the decision being taken. Further, it is multi-disciplinary, taking tools from subjects such as mathematics, statistics, engineering, economics, and intelligence and knowledge analysis to obtain possible alternative actions.

Therefore, we can define operations research in program management as follows:

- Scientific, analytical, experimental, and quantitative methodology which, by assessing the overall implication of various alternative courses of action in a management system, provides a systematic basis for decisions regarding the operations under the decision maker's control and eventually an improved basis for management decisions
- A decision-making method that uses scientific, mathematical, or logical means to attempt to cope with the problems that confront the program manager when aiming to achieve the right methodology for decision making in dealing with problems
- A methodology that optimizes both the design and the construction of different projects within a program, usually requiring the allocation of required resources to meet the completion date within the costs and specifications

- A basis for planned and scientific knowledge through an interdisciplinary approach in order to represent complex functional relationships as mathematical models for the purpose of providing a quantitative analysis to determine the best utilization of limited resources (Heiman 1987).

## Operations Research Mechanism

Given that operations research represents an integrated framework to help make decisions, it is important to have a clear understanding of this framework so that it can be applied to a generic problem. To achieve this, operations research mechanism comprises the following seven sequential steps: (1) orientation, (2) problem definition, (3) data collection, (4) model formulation, (5) solution, (6) model validation and output analysis, and (7) implementation and monitoring.

Figure 5.1 shows the operations research mechanism in program management and how to relate each step in a mechanism for continuous feedback.

## Operations Research Approach

Most operations research studies involve the construction of a mathematical model. The model is a collection of logical and mathematical relationships that represents aspects of the situation under study. Models describe important relationships between the variables, including an objective function with which alternative

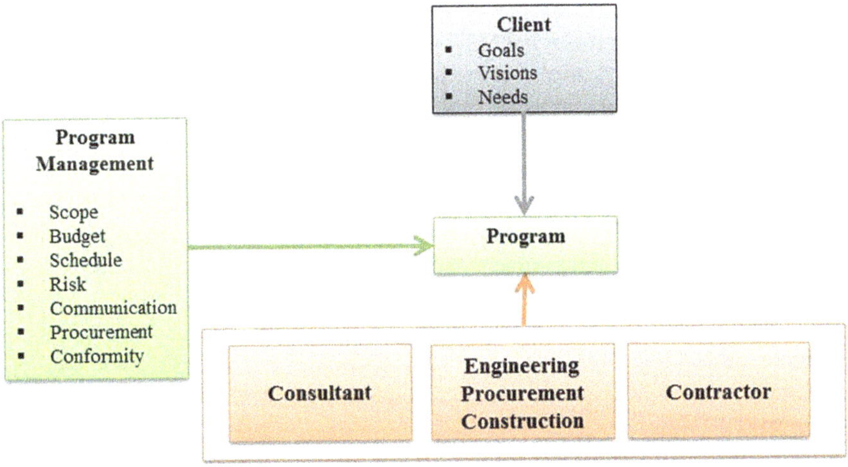


**Fig. 5.1** Operations research mechanism in program management

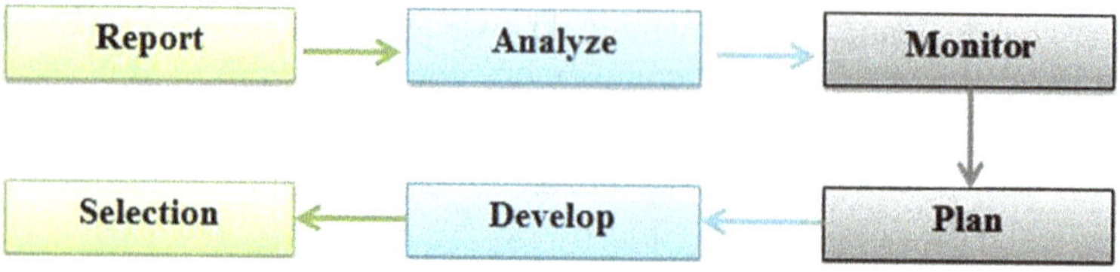


Fig. 5.2 Summary of operations research approach

solutions are evaluated, and constraints that restrict solutions to feasible values. Figure 5.2 shows the summary of operations research approach.

Although the analyst would hope to study the broad implications of the problem using a system approach, a model cannot include every aspect of a situation. A model is always an abstraction that is, of necessity, simpler than the real situation. Elements that are irrelevant or unimportant to the problem are ignored, hopefully leaving sufficient detail so that the solution obtained with the model has value with regard to the original problem.

A model must be representative of the original situation, capable of being solved, and valid. These objectives are often contradictory and are not always attainable. It is generally true that the most powerful solution methods can be applied to the simplest, or most abstract, model (Eiselt and Sandblom 2012).

The significant features of operations research include the following:

*Decision making*: Operations research aims to help the program manager to obtain an optimal solution with the use of mathematical, quantitative, statistical, and intelligent techniques. It also assists the decision maker to improve his creative and judicious capabilities and to analyze and understand the problem, thus leading to better control, better coordination, better systems, and finally better decisions.

*Scientific approach*: Operations research applies scientific methods, techniques, and tools for the purpose of analyzing and solving complex problems. In this approach, there is no place for guesswork or the human bias of the decision maker.

*Interdisciplinary team approach*: Basically, a program management problem is of a complex nature and, therefore, requires a team effort to handle it. This team comprises scientists/mathematicians and managers who jointly use the operations research tools to obtain an optimal solution to the problem. The program manager tries to analyze the cause-and-effect relationship between various parameters of the problem and evaluates the outcome of various alternative strategies.

*System approach*: The main aim of the system approach is to trace, for each proposal, all significant and indirect effects on a system and to evaluate each action in terms of effects on the system as a whole. The interrelationship and interaction of each subsystem can be handled with the help of mathematical/analytical models of operations research to obtain an acceptable solution.

The various steps required for the analysis of a problem under operations research are as follows:

1. *Observe the problem environment.* The first step of operations research study is to observe the environment in which the problem exists. The activities that constitute this step are the information, the knowledge, the parameters, the data, and the different components needed for the system. With the help of such activities, the operations research analyst is better prepared to formulate the problem.
2. *Analyze and define the problem.* In this step, the system to be analyzed and optimized will also use objectives and limitations in light of the problem. The end results of this step are a clear grasp of the need for a solution and an understanding of its nature.
3. *Develop a model.* Operations research models are basically mathematical models representing systems, process, or environment in the form of equations, relationships, or formulae. The activities in this step are to define interrelationships among variables, formulating equations, and using known operations research models or searching for suitable alternative models.
4. *Select your data input.* It is an established fact that, without authentic and appropriate data, the results of the operations research models cannot be trusted. Hence, obtaining the right kind of data is a vital step in the operations research process. Important activities in this step are to analyze internal/external data and facts, collect opinions, and use computer databases. The purpose of this step is to have sufficient input to operate and test the model.
5. *Solve and test.* The solution of the problem is obtained with the assistance of the model and the data input. Such a solution is not implemented immediately, but is used to test the model and to find its limitations, if any. If the solution is not reasonable or if the model is not behaving properly, updating and modification of the model are considered at this stage. The end result of this step is a solution that is desirable and supports current organizational objectives.
6. *Implement the solution.* In operations research, the decision making is scientific but implementing the decision involves many behavioral issues. Therefore, the implementation authority has to resolve these issues, which involve real site data, procurement, cost control, and different project teams such as the engineers, supporting staff, and project managers, to avoid further conflicts.

## General Mathematical Models

Mathematical models are a main part of operations research. They vary widely, and many programming aids, mathematical models, techniques, and algorithms are available. Linear programming, which is most commonly applied in the construction industry, is a typical mathematical program which consists of a single objective function, representing either a profit to be maximized or a cost to be minimized and a set of constraints that circumscribe the decision variables. This technique is the onc most widely used, researched, and applied in the construction industry. Linear programming is discussed in detail in the rest of this chapter, including how to formulate and solve basic problems using a linear program. In the case of a linear program, the objective function and constraints are all linear functions of the decision variables. At first glance, these restrictions would seem to limit the scope of the linear program model, but this is hardly the case. Because of its simplicity, software has been developed that is capable of solving problems containing millions of variables and tens of thousands of constraints. Countless real-world applications have been successfully modeled and solved using linear programming techniques. Linear programming has proven to be an extremely powerful tool, both in modeling real-world problems and as a widely applicable mathematical theory (Keil 2008).

However, many interesting optimization problems are nonlinear. The study of such problems involves a diverse blend of linear algebra, multivariate calculus, numerical analysis, and computing techniques. Important areas include the design of computational algorithms (including interior point techniques for linear programming), the geometry and analysis of convex sets and functions, and the study of specially structured problems such as quadratic programming. Nonlinear optimization provides fundamental insights into mathematical analysis and is widely used in a variety of fields such as engineering design, regression analysis, inventory control, geophysical exploration, and economics (Chinneck et al. 2000).

Other techniques that can be used in practical decisions and provide optimal solutions for construction programs are listed below. These are mentioned as for reference and are summarized in the following sections.

## Network Flow Programming

The term network flow program describes a type of model that is a special case of the more general linear program. The class of network flow programs includes such problems as the transportation problem, the assignment problem, the shortest path problem, the maximum flow problem, the pure minimum cost flow problem, and the generalized minimum cost flow problem. It is an important class because many aspects of actual situations are readily recognized as networks and the representation of the model is much more compact than the general linear program. When a situation can be entirely modeled as a network, very efficient algorithms exist for the solution of the optimization problem, and network flow programming is many times more efficient than linear programming in the utilization of computer time and space resources.

## Integer Programming

Integer programming is concerned with optimization problems in which some of the variables are required to take on discrete values. Rather than allowing a variable to assume all real values in a given range, only predetermined discrete values within the range are permitted. In most cases, these values are the integers, giving rise to the name of this class of models. Models with integer variables are very useful. Situations that cannot be modeled by linear programming are easily handled by integer programming. Primary among these involve binary decisions such as yes–no, build–no build, or invest–not invest. Although one can model a binary decision in linear programming with a variable that ranges between 0 and 1, there is nothing that keeps the solution from obtaining a fractional value such as 0.5, which is hardly acceptable to a decision maker. Integer programming requires such a variable to be either 0 or 1, but not in-between.

Unfortunately, integer programming models of practical size are often very difficult or impossible to solve. Linear programming methods can solve problems that are orders of magnitude larger than those solvable by integer programming methods. Still, many interesting problems are solvable, and the growing power of computers makes this an active area of interest in operations research.

## Nonlinear Programming

When expressions defining the objective function or constraints of an optimization model are not linear, one has a nonlinear programming model. Again, the class of situations appropriate for nonlinear programming is much larger than the class for linear programming. Indeed, it can be argued that all linear expressions are really approximations for nonlinear ones. Since nonlinear functions can assume such a wide variety of functional forms, there are many different classes of nonlinear programming models. The specific form has much to do with how easily the problem is to solve, but in general, a nonlinear programming model is much more difficult to solve than a similarly sized linear programming model.

## Dynamic Programming

Dynamic programming models are represented in a different way to other mathematical programming models. Rather than an objective function and constraints, a dynamic programming model describes a process in terms of states, decisions, transitions, and returns. The process begins in some initial state where a decision is made. This decision causes a transition to a new state. Based on the starting state, ending state, and decision, a return is realized. The process continues through a sequence of states until finally a final state is reached. The problem is to find the sequence that maximizes the total return.

Although traditional integer programming problems can be solved with dynamic programming, the models and methods are most appropriate for situations that are not easily modeled using the constructs of mathematical programming. Objectives with very general functional forms may be handled, and a global optimal solution is always obtained. The price of this generality is computational effort. Solutions to practical problems grow exponentially with the number of dimensions of the problem. While, generally, dynamic programming is capable of solving many diverse problems, it may require huge computer storage in most cases.

## Stochastic Programming

The mathematical programming models, such as linear programming, network flow programming, and integer programming, generally neglect the effects of uncertainty and assume that the results of decisions are predictable and deterministic. This abstraction of reality allows large and complex decision problems to be modeled and solved using powerful computational methods. Stochastic programming explicitly recognizes uncertainty by using random variables for some aspects of the problem. With probability distributions assigned to the random variables, an expression can be written for the expected value of the objective to be optimized. Then, a variety of computational methods can be used to maximize or minimize the expected value.

## Combinatorial Programming

The most general type of optimization problem, and one that is applicable to most spreadsheet models, is the combinatorial optimization problem. Many spreadsheet models contain variables and compute measures of effectiveness. The spreadsheet user often changes the variables in an unstructured way to look for the solution that obtains the greatest or least of the measure. In the words of operations research, the analyst is searching for the solution that optimizes an objective function, the measure of effectiveness. Combinatorial optimization provides tools for automating the search for good solutions and can be of great value for spreadsheet applications.

## Simulation

When a situation is affected by random variables, it is often difficult to obtain closed-form equations that can be used for evaluation. Simulation is a very general technique for estimating statistical measures of complex systems. A system is modeled as if the random variables were known. Then, values for the variables are drawn randomly from their known probability distributions. Each replication gives one observation of the system response. By simulating a system in this fashion over many replications and recording the responses, one can compute statistics concerning the results. The statistics are used for evaluation and design.

## Constraint Satisfaction

Many industrial decision problems involving continuous constraints can be modeled as continuous constraint satisfaction and optimization problems. Constraint satisfaction problems are large in size and in most cases involve transcendental functions. They are widely used in modeling and optimization of cost restrictions.

## Convex Program

The term convex program covers a broad class of optimization problems. When the objective function is convex and the feasible region is a convex set, both of these assumptions are enough to ensure that the local minimum is a global minimum.

## Heuristic Optimization

Several heuristic tools have evolved that facilitate solving optimization problems that were previously difficult or impossible to solve. These tools include evolutionary computation, simulated annealing, and particle swarm, etc. Common approaches include, but are not limited to:

- Comparing solution quality to optimum on benchmark problems with known optima, average difference from optimum, and frequency with which the heuristic finds the optimum
- Comparing solution quality to a best known bound for benchmark problems whose optimal solutions cannot be determined
- Comparing the achieved heuristic to published heuristics for the same problem type, difference in solution quality for a given run time and, if relevant, memory limit
- Profiling average solution quality as a function of run time; for instance, plotting mean and either min. and max. or 5th and 95th percentiles of solution value as a function of time—this assumes that one has many benchmark problem instances that are comparable (Aronson et al. 2008; Rothlauf 2011).

## Optimization Techniques and Mathematical Programming

Mathematical optimization is the branch of computational science that seeks to find the optimal solutions to problems in which the quality of any answer or solution can be expressed as a numerical value. Such problems arise very often in program management in the areas of allocating resources, minimizing costs, and optimizing the time in the performance and execution of projects. The range of techniques available to solve them is nearly as wide.

A mathematical optimization model consists of an objective function and a set of constraints expressed in the form of a system of equations or inequalities. Optimization models are used extensively in almost all areas of decision making, such as engineering design and financial portfolio selection. This part of the chapter presents a focused and structured process for linear programming and its applications to different aspects of program management.

If the mathematical model is a valid representation of the performance of the system, and if the appropriate analytical techniques are applied, then the solution obtained from the model should also be the solution to the system problem. The effectiveness of the results of the application of any optimization technique is largely a function of the degree to which the model represents the system studied (Belegundu and Arora 2005).

To define those conditions that will lead to the solution of a system problem, the analyst must first identify a criterion by which the performance of the system may be measured. This criterion is often referred to as the measure of the system performance or the measure of effectiveness. The mathematical (i.e., analytical) model that describes the behavior of the measure of effectiveness is called the objective function. If the objective function is to describe the behavior of the measure of effectiveness, it must capture the relationship between that measure and those variables that cause it to change. System variables can be categorized as decision variables and parameters. A decision variable is a variable that can be directly controlled by the decision maker. There are also some parameters whose values might be uncertain for the decision maker. This calls for sensitivity analysis after the best strategy has been found. In practice, mathematical equations rarely capture the precise relationship between all system variables and the measure of effectiveness (Rothlauf 2011).

The general procedure for optimization techniques that can be used in the process cycle of modeling is to: (1) describe the problem; (2) prescribe a solution; and (3) control the problem by assessing and updating the optimal solution continuously, while changing the parameters and structure of the problem. Clearly, there are always feedback loops among these general steps:

1. *Mathematical formulation of the problem must be validated.* Therefore, as soon as the program manager detects a problem, he must understand it in order to adequately describe the problem in writing. He will, then, develop a mathematical model or framework to represent reality in order to devise and use an optimization solution algorithm. A good mathematical formulation for optimization must be both inclusive (i.e., it includes what belongs to the problem) and exclusive (i.e., shaving off what does not belong to the problem).
2. *Find an optimal solution.* This is an identification of a solution algorithm and its implementation stage.

3. *Post-solution analysis*. This activity includes updating the optimal solution in order to control the problem. It is crucial to periodically update the optimal solution to any given optimization problem. A model that was valid may lose validity due to changing conditions, thus becoming an inaccurate representation of reality and adversely affecting the ability of the decision maker to make good decisions. The optimization model created should be able to cope with changes.
4. *Heed the importance of feedback and control*. It is necessary to place heavy emphasis on the importance of thinking about the feedback and control aspects of an optimization problem. The very nature of the optimal strategy's environment is changing, and therefore, feedback and control are an important part of the optimization modeling process (Bradley et al. 1997).

Mathematical programming solves the problem of determining the optimal allocations of limited resources that are required to meet a given objective. The objective must represent the goal of the decision maker. For example, the resources may correspond to engineers, labor, equipment, materials, money, or land. Out of all permissible allocations of the resources, it is desired to find the one or ones that maximize or minimize some numerical quantity such as profit or cost.

Optimization problems are made up of three basic ingredients:

1. *An objective function*—that is, the quantity we want to minimize or maximize. Most optimization problems have a single objective function; if they do not, they can often be reformulated so that they do.
2. *Controllable inputs*—these are the set of decision variables that affect the value of the objective function. In a construction problem, the variables might include the allocation of different available resources, or the labor spent on each activity. Decision variables are essential. If there are no variables, we cannot define the objective function and the problem constraints.
3. *Uncontrollable inputs*—these are called parameters. The input values may be fixed numbers associated with the particular problem. We call these values parameters of the model. Often you will have several "cases" or variations of the same problem to solve, and the parameter values will change in each problem variation.
4. *Constraints* are relations between decision variables and the parameters. A set of constraints allows some of the decision variables to take on certain values and exclude others. For the construction problem, the concept of spending a negative amount of time on any activity does not make sense, so we constrain all the "time" variables to be nonnegative. Constraints are not always essential. In fact, the field of unconstrained optimization is a large and important one for which many algorithms and software are available. In practice, answers that make good sense about the underlying physical or economic problems cannot often be obtained without putting constraints on the decision variables.

A solution value for decision variables where all of the constraints are satisfied is called a feasible solution. Most solution algorithms proceed by first finding a feasible solution, then seeking to improve upon it, and finally changing the decision variables to move from one feasible solution to another. This process is repeated until the objective function has reached its maximum or minimum. This result is called an optimal solution.

## Linear Programming—An Introduction

The development of linear programming has saved many thousands or millions of dollars for most companies and businesses of even moderate size in the various industrialized countries of the world, and its use in other sectors of society has been spreading rapidly. A major proportion of optimization computations on computers is devoted to the use of linear programming. The most common type of application involves the general problem of allocating limited resources among competing activities in the best possible (i.e., optimal) way (Vanderbei 2013).

More precisely, this type of problem solving involves selecting the levels of certain activities that compete for scarce resources that are necessary to perform those activities. The choice of activity levels then dictates how much of each resource will be consumed by each activity. The one common ingredient in each of these situations is the necessity for allocating resources to activities by choosing the levels of those activities.

Linear programming uses a mathematical model to describe the problem of concern. The adjective "linear" means that all the mathematical functions in this model are required to be linear functions (Gass 2010).

The word programming does not refer here to computer programming; rather, it is essentially a synonym for planning. Thus, linear programming involves the planning of activities to obtain an optimal result, i.e., a result that reaches the specified goal best (according to the mathematical model) among all feasible alternatives. Although allocating resources to activities is the most common type of application, linear programming has numerous other important applications as well. In fact, any problem whose mathematical model fits the very general format for the linear programming model is a linear programming problem. These are some of the reasons for the tremendous impact of linear programming in recent decades.

The ability to introduce linear programming using a graphical approach, the relative ease of the solution method, the widespread availability of linear programming software packages, and the wide range of applications make linear programming accessible to a wide range of professionals in the construction industry generally. Additionally, linear programming provides an excellent opportunity to introduce the idea of "what-if" analysis, due to the powerful tools for post-optimality analysis developed for the linear programming model (Chvatal 1983).

Linear programming deals with a class of problems where both the objective function to be optimized is linear and all relations among the variables corresponding to resources are linear. They include the following:

1. *Resource allocation* such as engineers, staff, foremen, workers, and equipment. This can be done by the optimal distribution of the resources between the different projects of the program.
2. *Cash distribution and procurement methods.* Procurement and availability of cash can be optimized to accelerate certain projects in delay and slow down others if not critical.
3. *Material distributions to site.* This is important if certain resources are not available, scarce, or not in stock. This is also critical if the market is in need of certain materials in peak times. Cement, steel, blocks, plumbing materials, ceramics, and others can be purchased by optimizing their use and need to the program. This factor can be crucial for a large program or the construction of a program in third-world countries.
4. *Optimization of the program* using the Critical Path Method (CPM), crashing certain activities to achieve the optimal completion date.

Any linear program consists of four parts: a set of decision variables, the parameters, the objective function, and a set of constraints. In formulating a given decision problem in mathematical form, the user should practice understanding the problem by carefully reading and rereading the problem statement (Dantzig 1998).

## Linear Programming—Problem Formulation

A linear programming problem, therefore, has three components:

1. A set of nonnegative variables, called decision variables, $x_j \geq 0$ for all $j$
2. An objective function to be maximized or minimized as the case may be. Let $z$ be the objective function, then maximize (or minimize)

$$z = c_1x_1 + c_2x_2 + c_3x_3 + \cdots + c_nx_n \tag{5.1}$$

3. A set of linear constraints, which form a system of equations or in-equations

$$a_{11}x_{11} + a_{12}x_2 + \cdots + c_{1n}x_n \geq \text{or} = \text{or} \leq b_1 \tag{5.2}$$

$$\begin{array}{cccccc} a_{21}x_{11}+ & a_{22}x_2+ & \cdots & +c_{2n}x_n & \geq \text{or} = \text{or} \leq b_2 \\ \vdots & \vdots & \vdots & \vdots & \vdots \\ a_{m1}x_{11}+ & a_{m2}x_2+ & \cdots & +c_{mn}x_n & \geq \text{or} = \text{or} \leq b_{m2} \end{array} \tag{5.3}$$

The problem takes, then, the form of—
Maximize or minimize

$$z = \sum_{j=1}^{n} c_jx_j \text{ subject to} \tag{5.4}$$

$$\sum_{j=1}^{n} a_{ij}x_j \geq \text{or} = \text{or} \leq b_i \sum \qquad (5.5)$$
$$x_j \geq 0$$

where:

$x_j$ = level of activity/activity type
$c_j$ = worth of one unit of activity $j$
$\sum_{j=1}^{n} a_{ij}x_j \geq \text{or} = \text{or} \leq b_i$ = availability of resource $i$
$a_{ij}$ is the number of unit (amount) of resource $i$ to produce 1 unit of activity $j$.

In the linear programming problem formulation above, the expression being optimized is called the objective function ($z$).

The variables $x_1, x_2, \ldots, x_n$ are called decision variables, and their values are subject to $m + 1$ constraints (every line ending with a $b_i$, plus the nonnegativity constraint).

A set of $x_1, x_2, \ldots, x_n$ satisfying all the constraints is called a feasible point, and the set of all such points is called the feasible region.

The solution of the linear program must be a point $(x_1, x_2, \ldots, x_n)$ in the feasible region, or else not all the constraints would be satisfied.

Not all linear programming problems are so easily solved. There may be many variables and many constraints. Some variables may be constrained to be nonnegative and others unconstrained. Some of the main constraints may be equalities and others inequalities. However, two classes of problems, called here, the standard maximum problem and the standard minimum problem, play a special role. In these problems, all variables are constrained to be nonnegative, and all main constraints are inequalities (Bertsimas et al. 1997; Schrijver 1998; Matousek and Gartner 2007).

## Terminology

1. The function to be maximized or minimized is called the objective function.
2. A vector, $\boldsymbol{x}$ for the standard maximum problem or $\boldsymbol{y}$ for the standard minimum problem, is said to be feasible if it satisfies the corresponding constraints.
3. The set of feasible vectors is called the constraint set.
4. A linear programming problem is said to be feasible if the constraint set is not empty; otherwise, it is said to be infeasible.
5. A feasible maximum (resp. minimum) problem is said to be unbounded if the objective function can assume arbitrarily large positive (resp. negative) values at feasible vectors; otherwise, it is said to be bounded. Thus, there are three possibilities for a linear programming problem: It may be bounded feasible, it may be unbounded feasible, and it may be infeasible.

6. The valuc of a bounded feasible maximum (resp. minimum) problem is the maximum (resp. minimum) value of the objective function as the variables range over the constraint set.
7. A feasible vector at which the objective function achieves the value is called optimal.

## Assumptions

Before we get too focused on the techniques of solving linear programs, it is important to review some theory. For instance, several assumptions are implicit in linear programming problems. These assumptions are as follows:

1. *Proportionality*: The contribution of any variable to the objective function or constraints is proportional to that variable. This implies no discounts or economies to scale. For example, the value of $8x_1$ is twice the value of $4x_1$, no more or less.
2. *Additivity*: The contribution of any variable to the objective function or constraints is independent of the values of the other variables.
3. *Divisibility*: Decision variables can be fractions. However, by using a special technique called integer programming, we can bypass this condition. Integer programming is beyond the scope of this book.
4. *Certainty*: This assumption is also called the deterministic assumption. This means that all parameters (all coefficients in the objective function and the constraints) are known with certainty. Realistically, however, coefficients and parameters are often the result of guesswork and approximation. The effect of changing these numbers can be determined with sensitivity analysis, which will be explored later in this chapter (Dantzig 1998).

## Basic Transformations

1. Minimizing $f(x)$ is equivalent to maximizing the negative worth of $f(x)$.

$$\text{Minimize} f(x) \equiv g(x), \quad \text{where } g(x) = -f(x) \tag{5.6}$$

2. An inequality in one direction can be changed to an inequality in the opposite direction by multiplying both sides of the inequality by $(-1)$.

$$a_1x_1 + a_2x_2 \geq b - a_1x_1 - a_{12}x_2 \leq b \tag{5.7}$$

3. An equation may be replaced by two inequalities in opposite directions.

$$a_1x_1 + a_2x_2 = b \tag{5.8}$$

is equivalent to

$$a_1x_1 + a_2x_2 \le b \quad \text{and} \quad a_1x_1 + a_2x_2 \ge b \text{ and similarly} \tag{5.9}$$

$$a_1x_1 + a_2x_2 \le -b \quad \text{and} \quad -a_1x_1 - a_2x_2 \le -b \tag{5.10}$$

Linear programming problems can face difficulties and will require special techniques in formulation. There may be many variables and many constraints. Some variables may be constrained to be nonnegative and others unconstrained. Some of the main constraints may be equalities and others inequalities. However, two classes of problems, called here the standard maximum problem and the standard minimum problem, play a special role. In these problems, all variables are constrained to be nonnegative, and all main constraints are inequalities (Chvatal 1983).

The standard minimum problem: Find an $m$-vector, $y = (y_1, \ldots, y_{m)}$, to minimize

$$y^Tb = y_1b_1 + \cdots + y_mb_m \tag{5.11}$$

Subject to the constraints

$$y_1a_{11} + y_2a_{21} + \cdots + y_ma_{m1} \ge c_1 \tag{5.12}$$

$$y_1a_{12} + y_2a_{22} + \cdots + y_ma_{m2} \ge c_2 \quad (\text{or } \; y^TA \ge c^T) \tag{5.13}$$

$$y_1 \ge 0, y_2 \ge 0, \ldots y_m \ge 0 \tag{5.14}$$

and

$$y_1a_{1n} + y_2a_{2n} + \cdots + y_ma_{mn} \ge c_n \quad (\text{or } y \ge 0) \tag{5.15}$$

Note that the main constraints are written as ≤ for the standard maximum problem and ≥ for the standard minimum problem. The introductory example is a standard maximum problem.

## Graphical Solution for Linear Programming—An Example

The following example illustrates that geometrically interpreting the feasible region is a useful tool for solving linear programming problems with two decision variables. The linear program is as follows:

$$\begin{array}{ll} \text{Minimize} & 4x_1 + x_2 = z \\ \text{Subject to} & 3x_1 + x_2 \geq 10 \\ & x_1 + x_2 \geq 5 \\ & x_1 \geq 3 \\ & x_1, x_2 \geq 0. \end{array} \tag{5.16}$$

The system of inequalities is plotted as the shaded region in Fig. 5.3. Since all of the constraints are greater than or equal to constraints, the shaded region above all three lines is the feasible region. The solution to this linear program must lie within the shaded region.

Recall that the solution is a point $(x_1; x_2)$ such that the value of $z$ is the smallest it can be, while still lying in the feasible region. Since $z = 4x_1 + x_2$, plotting the line $x_1 = (z - x_2)/4$ for various values of $z$ results in isocost lines, which have the same slope. Along these lines, the value of $z$ is constant. In Fig. 5.3, the dotted lines represent isocost lines for different values of $z$. Since isocost lines are parallel to each other, the thick dotted isocost line for which $z$ = 14 is clearly the line that

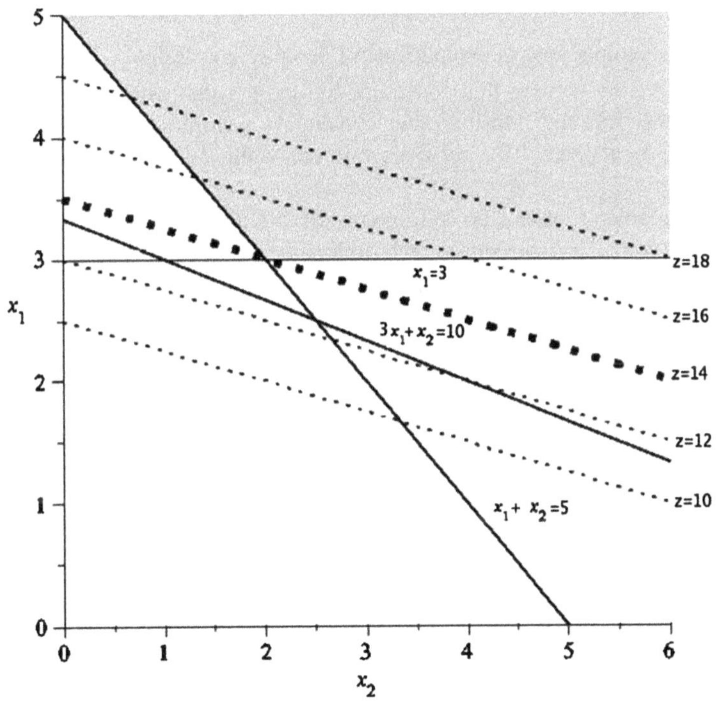


**Fig. 5.3** The shaded region above all three solid lines is the feasible region (one of the constraints does not contribute to defining the feasible region). The dotted lines are isocost lines. The thick isocost line that passes through the intersection of the two defining constraints represents the minimum possible value of $z$ = 14 while still passing through the feasible region

intersects the feasible region at the smallest possible value for $z$. Therefore, $z = 14$ is the smallest possible value of $z$ given the constraints. This value occurs at the intersection of the lines $x_1 = 3$ and $x_1 + x_2 = 5$, where $x_1 = 3$ and $x_2 = 2$.

## A Standard Form for all Linear Programming Problems

A linear programming problem was defined as maximizing or minimizing a linear function subject to linear constraints. All such problems can be converted into the form of a standard maximum problem by the following techniques.

A minimum problem can be changed to a maximum problem by multiplying the objective function by −1. Similarly, constraints of the form can be changed into the form

$$\sum_{j=1}^{n} = 1(-a_{ij})x_j \leq -b_i \quad \sum_{j=1}^{n} a_{ij}x_j \geq b_i \tag{5.17}$$

Two other problems arise.

1. Some constraints may be equalities. An equality constraint $\sum_{j=1}^{n} a_{ij}x_j \geq b_i$ may be removed by solving this constraint for some $x_j$ for which $a_{ij} \neq 0$ and submitting this solution into the other constraints and into the objective function wherever $x_j$ appears. This removes one constraint and one variable from the problem.
2. Some variables may not be restricted to be nonnegative. An unrestricted variable, $x_j$, may be replaced by the difference of two nonnegative variables, $x_j = u_j - v_j$, where $u_j \geq 0$ and $v_j \geq 0$. This adds one variable and two nonnegativity constraints to the problem.

Any theory derived for problems in standard form is therefore applicable to general problems. However, from a computational point of view, the enlargement of the number of variables and constraints in (5.2) is undesirable and, as will be seen later, can be avoided.

## Linear Programming Practical Examples in Construction

Having learnt the theory behind linear programs, we will focus in this section on practical methods of solving them in construction.

## *Concrete Batches Problem*

Suppose that there are $m$ batches that generate concrete and $n$ sites. The amount of concrete generated at source $i$ is $a_{i,}$ and the requirement of site $j$ is $b_j$. It is desired to select appropriate transfer facilities from among $K$ candidate facilities. Potential transfer facility $k$ has fixed cost $f_k$, capacity $q_{k,}$ and unit processing cost $\alpha_k$ per ton of concrete. Let $c_{ik}$ and $c_{kj}$ be the unit transport costs from batch $i$ to transfer station $k$ and from transfer station $k$ to site $j$, respectively. The problem is to choose the transfer facilities and the transportation pattern that minimize the total capital and operating costs of the transfer stations plus the transportation costs.

Let:

$w_{ik}$ = tons of concrete moved from $i$ to $k$, $1 \leq i \leq m$, $1 \leq k \leq K$
$y_k$ = a binary variable that equals 1 when transfer station $k$ is used and 0 otherwise,

$$1 \leq k \leq K$$

$x_{kj}$ = tons of concrete moved from $k$ to $j$, $1 \leq k \leq K$, $1 \leq j \leq n$

The objective function features several double sums in order to describe all the costs faced in the process of concrete distribution. The objective function is as follows:

$$\begin{aligned} \min z = &\sum_i \sum_k c_{ik} w_{ik} + \sum_k \sum_j c_{kj} w_{kj} \\ &+ \sum_k f_k y_k + \sum_k \sum_i \alpha_k w_{ik} \end{aligned} \tag{5.18}$$

The first constraint equates the tons of concrete coming from all the sources with the tons of concrete going to all the sites. This constraint is as follows:

$$\sum_i w_{ik} = \sum_i x_{kj}. \tag{5.19}$$

The next constraint says that the amount of concrete produced equals the amount moved to all the transfer facilities:

$$\sum_k w_{ik} = \sum_i \alpha_i. \tag{5.20}$$

Next, there must be a constraint that restricts how much concrete is at transfer or at sites, depending on their capacity. This restriction gives:

$$\sum_k x_{kj} = b_j \quad \text{and} \quad \sum_i w_{ik} \leq q_k \tag{5.21}$$

Putting these constraints all together, the linear program is as follows:

$$
\begin{aligned}
\text{Minimize} \quad & z = \sum_i \sum_k c_{ik} w_{ik} + \sum_k \sum_j c_{kj} w_{kj} + \sum_k f_k y_k + \sum_k \sum_i \alpha_k w_{ik} \\
\text{Subject to} \quad & \sum_k w_{ik} = \sum_j x_{kj} \\
& \sum_k w_{ik} = \sum_i a_i \\
& \sum_k x_{kj} \leq b_j \\
& \sum_i w_{ik} \leq q_k
\end{aligned} \tag{5.22}
$$

$$\text{All variables} \geq 0. \tag{5.23}$$

Most linear program-solving software allows the user to designate certain variables as binary, so ensuring this property of $y_k$ would not be an obstacle in solving this problem.

## *Resources Problem*

Let $y_i$ be the number of units of resources $F_i$ to be purchased per month. The cost per month of resources is as follows:

$$b_1 y_1 + b_2 y_2 + \cdots + b_m y_m. \tag{5.24}$$

The amount of products $N_j$ contained in this resource is as follows:

$$a_{1j} y_1 + a_{2j} y_2 + \cdots + a_{mj} y_m \tag{5.25}$$

for $j = 1, \ldots, n$. We do not consider such a product unless all the minimum monthly requirements are met, that is, unless

$$a_{1j} y_1 + a_{2j} y_2 + \cdots + a_{mj} y_m \geq c_j \quad \text{for } j = 1, \ldots, n. \tag{5.26}$$

Of course, we cannot purchase a negative amount of resources, so we automatically have the constraints

$$y_1 \geq 0, y_2 \geq 0, \ldots, y_m \geq 0. \tag{5.27}$$

Our problem is as follows: minimize (5.1) subject to (5.2) and (5.3). This is exactly the standard minimum problem.

## The Transportation Problem

There are $I$ ports, or production plants, $P_1, \ldots, P_I$, that supply a certain commodity, and there are $J$ projects, $M_1, \ldots, M_J$, to which this commodity must be shipped. Port $P_i$ possesses an amount of the commodity ($i = 1, 2, \ldots, I$), and project $M_j$ must receive the amount $r_j$ of the commodity ($j = 1, \ldots, J$). Let $b_{ij}$ be the cost of transporting one unit of the commodity from port $P_i$ to project $M_j$. The problem is to meet the program requirements at minimum transportation cost.

Let $y_{ij}$ be the quantity of the commodity shipped from port $P_i$ to project $M_j$. The total transportation cost is as follows:

$$\sum_{i=1}^{I} \sum_{j=1}^{J} y_{ij} b_{ij}. \tag{5.28}$$

The amount sent from port $P_i$ is ${}_{j=1}\, y_{ij,}$ and since the amount available at port $P_i$ is $s_i$, we must have:

$$\sum_{j=1}^{J} y_{ij} \leq s_i \quad \text{for}\, i = 1, \ldots, I. \tag{5.29}$$

The amount sent to project $\text{M}_j$ is $M_j \sum_{i=1}^{I} y_{ij}$, and since the amount required there is $r_j$, we must have:

$$\sum_{i=1}^{I} y_{ij} \geq r_j \quad \text{for}\, j = 1, \ldots, J. \tag{5.30}$$

It is assumed that we cannot send a negative amount from $P_I$ to $M_j$, so we have:

$$y_{ij} \geq 0 \quad \text{for}\, i = 1, \ldots, I \quad \text{and} \quad j = 1, \ldots, J. \tag{5.31}$$

Our problem is as follows: minimize (5.4) subject to (5.5)–(5.7).

## Activity Analysis Problem

There are $n$ activities, $A_1, \ldots, A_n$, that a program manager may employ, using the available supply of $m$ resources, $R_1, \ldots, R_m$ (labor hours, steel, etc.). Let $b_i$ be the available supply of resource $R_i$. Let $a_{ij}$ be the amount of resource $R_i$ used in operating activity $A_j$ at unit intensity. Let $c_j$ be the net value to the program manager of operating activity $A_j$ at unit intensity. The problem is to choose the intensities at which the various activities are to be operated to maximize the value of the output to the company subject to the given resources.

Let $x_j$ be the intensity at which $A_j$ is to be operated. The value of such an activity allocation is as follows:

$$\sum_{j=1}^{n} c_j x_j. \tag{5.32}$$

The amount of resource $R_i$ used in this activity allocation must be no greater than the supply, $b_{i;}$ that is:

$$\sum_{j=1}^{n} a_{ij} x_j \leq b_i \quad \text{for } i = 1, \ldots, m. \tag{5.33}$$

It is assumed that we cannot operate an activity at negative intensity, that is:

$$x_1 \geq 0, x_2 \geq 0, \ldots, x_n \geq 0. \tag{5.34}$$

Our problem is as follows: maximize (5.8) subject to (5.9) and (5.10). This is exactly the standard maximum problem.

Construction management involves the coordination of a group of activities, whereas the manager plans, organizes, employs, directs, and controls construction projects and programs to achieve an object, including a specification of their interrelationships and considering the required resources in an acceptable time span. The success of the Critical Path Method (as described in Chap. 3) is that it utilizes the planner's knowledge, experience, and instincts in a logical way, first to plan and then to schedule. Critical Path Method can save money through better planning.

## *Linear Programming Technique to Find the Critical Path for the Program Network*

Program managers are continually facing a situation in which they must take a decision whether to complete a project, or a series of projects, sooner than originally specified in the contract in order to accelerate the completion of the program or utilize the resources of this project on other projects yet to start. They also need to optimize the cost of expediting the program.

One of the most challenging jobs that any program manager can take on is the management of a large-scale program that requires coordinating numerous activities throughout the organization. A myriad of details must be considered in planning how to coordinate all these activities, develop a realistic schedule, and then monitor the progress of the program.

This example is to find the duration of program completion, i.e., how long it takes to complete the program (critical path or longest route of a unit flow entering at the start node and terminating at the finish node). The model is built because sometimes it is required to complete a project within a program within the predetermined deadline to keep the cost at its lowest possible level. Failure to do so ultimately leads to an increase in the total cost of the project and ultimately the program.

Before formulating the model, let us define some relevant terms. To simplify this problem, we know that a program is the combination of some activities which are, in this instance, the different projects. As factors such as design, procurement, and technical and logistic support are not included, the projects are interrelated in a logical sequence in the sense that the starting of some projects is dependent upon the completion of some other projects. These projects are activities $A_i$, which require time and resources to be completed. The relationship between the activities is specified in terms of events. An event represents a point in time that implies the completion of some activities and the beginning of new ones. The beginning and end point of an activity are thus expressed by two events.

Now let us define the variables of the problem.

$Y_i$ = The decision variable for the start time of project $I$; the time when an event $i$ will occur, measured since the beginning of the program, where $i = (1, 2, 3, \ldots, n)$
$A_i$ = Program activities, where $i = (1, 2, \ldots, n)$. These are the projects
$T_N$, $i$ = Normal time for project $i$, which is usually the time to complete the project with minimal resource

The objective is to minimize the cost of crashing the total program by minimizing the durations of crashing projects multiplied by their associated costs slope, then adding the resultant cost to the normal cost of program completion.

We know linear programming is a tool for decision making under certain situations. So the basic assumption of this approach is that we have to know some relevant data with certainty. We are interested in finding the critical path or the longest route of a unit flow entering at the start node and terminating at the finish node.

The following terms are needed to illustrate this procedure.

Thus, the objective function of linear programming becomes:

$$\text{Maximize}\,(Z) = \sum_{i=1}^{n} T_{\mathrm{N},i} \cdot Y_i \tag{5.35}$$

This objective function is subject to some constraints, which can be classified into three categories:

- For activities that enter node $i$

$$\sum_{i=1}^{n} Y_{i\cdot} = 1 \tag{5.36}$$

- For activities that leave node $i$

$$\sum_{i=1}^{n} Y_{i\cdot} = -1 \tag{5.37}$$

- For activities that enter and leave node $i$:

For each node, there is one constraint that represents the conservation of flow: total input flow = total output flow.

In this formulation, the $Y_i = 0$ or 1 denotes the absence or presence of unit flow from one node to another. So

$$\sum_{i=1}^{n} +Y - Y_1 = 0 \tag{5.38}$$

- Nonnegative constraints:

All the decision variables $Y_i \geq 0$

## Applications of Linear Programming in Program Management

In this section, some applications of linear programming in program management are described to highlight the importance of linear programming in this field.

### *A Linear Programming Approach for Project Control*

One of the most challenging jobs that any manager can take on is the management of a large-scale project or program that requires coordinating numerous activities throughout the organization. A myriad of details must be considered in planning how to coordinate all these activities, develop a realistic schedule, and then monitor the progress of the project.

The main aim of using the linear programming technique for project control is to build two models. The first is to find how long it takes to complete the program (critical path or longest route of a unit flow entering at the start node and terminating at the finish node). The second model is built because sometimes it is required to complete a program within the predetermined deadline to keep cost at its lowest possible level. Failure to do so ultimately leads to increase in total cost. This would lead managers to encounter a decision situation in which some activities of the program will be crashed to minimize the total cost of crashing. Thus, the second model is to minimize the cost of crashing the program's activities to meet the desired program completion time and to deal with all the data needed to develop a schedule information system and then to monitor the progress of the program. Finally, an analysis of the results obtained from solving these two models will be required to give some flexibility in planning, scheduling, and controlling.

## *Linear Programming Approach to Optimizing Strategic Investment in the Construction Workforce*

The construction industry has been facing several challenges, including a shortage of skilled workers. One of the key reasons for this problem is the absence of human resource management strategies for construction workers at project, corporate, regional, or program levels. Linear programming is a useful tool to address the issues of workforce training and allocation on a construction program. The objective of the model is to minimize labor costs while satisfying the program labor demands. It presents a framework to optimize the investment in, and to make the best use of, the available workforce with the intent to reduce the program costs and improve schedule performance.

A linear program model is required to provide an optimization-based framework for matching supply and demand of construction labor most efficiently through training, recruitment, and allocation. Given a program schedule or demand profile and the available pool of workers, the suggested model provides human resource managers with a combined strategy for training the available workers and hiring additional workers. The input data to the proposed model consist of a certain available labor pool, cost figures for training workers in different skills, the cost of hiring workers, hourly labor wages, and estimates of affinities between the different skills considered.

## *Optimizing Scheduling Programs Using Linear Programming*

Program management is the process of the application of knowledge, skills, tools, and techniques to program activities to meet program requirements. That is to say, program management is an interrelated group of processes that enables the program manager to achieve a successful program.

The functions of program management include the following:

(a) Planning—Planning the program and establishing its life cycle
(b) Organizing—Organizing resources: personnel, equipment, materials, facilities, and finances

(c) Leading—Assigning the right people to the right job, motivating people, and setting the course and goals for the program
(d) Controlling—Evaluating progress of program and, when necessary, get it back on track. Performing these functions in an organized framework of processes is the job of the program manager.

The program manager is assigned by an organization the responsibility and authority to manage a program, and the three basic objectives for which he is responsible for are as follows:

(a) Deliver a program that meets the requirements of the specification.
(b) Deliver a program that meets the requirements of the contract delivery schedule.
(c) Meet the company's profit objectives for the contract.

From a program management perspective, a program is considered a success if:

(a) The resulting construction delivery is delivered "within the specifications required"
(b) The system is delivered "on time"
(c) The system is delivered "within budget".

Program management has evolved as a new field with the development of two analytical techniques for planning, scheduling, and controlling of programs. These are the CPM (Chap. 3) and the Program Evaluation and Review Technique (PERT). These techniques are neither suitable nor adequate for addressing typical challenges related to time–cost trade-off. Optimal schedule cost can be determined by trial and error for small programs, but in a realistic program consisting of many activities, such trial and error becomes extremely tedious and impossible. A very limited number of computer programs are available but far from perfect. Such programs have a limited capacity to accept time–cost data. Thus, linear programming, as an optimization technique, has been developed to aid in the quick determination of the minimum cost for every possible value of program duration. Clearly, the use of this optimization techniques incorporated with time–cost trade-off becomes an economic necessity.

The models developed represent many restrictions and management considerations of the CPM and PERT. They could be used by program or project managers at a planning stage to explore numerous possible opportunities to the client and stakeholders, and predict the effect of a decision on the construction in order to facilitate a preferred operating policy given different management objectives.

An implementation using these methods is shown by many researchers to outperform several other techniques and a large class of test problems. Linear programming shows that the algorithm is very promising in practice on a wide variety of time–cost crash problems. Theses methods are simple, applicable to a large network, and generate a shorter computational time at lower cost, along with an increase in robustness (Elmabrouk 2012).

# References


Aronson, J. E., & Zionts, S. (2008). *Operations research: Methods, models, and applications*. Paperback, April 28.

Belegundu, A. D., & Arora, J. S. (2005). A study of mathematical programming methods for structural optimization. Part I: Theory. *International Journal for Numerical Methods in Engineering, 21*(9), 1583–1599. doi:10.1002/nme.1620210904.

Bertsimas, D., & Tsitsiklis, J. N. (1997). Introduction to linear optimization. *Athena Scientific Series in Optimization and Neural Computation, 6*, Hardcover, February 1.

Bradley, S., Hax, A., & Magnanti, T. (1997). *Applied mathematical programming*. Reading, MA: Addison-Wesley Publishing Company.

Chinneck, J. W., Cha, P. D., Rosenberg, J. J., & Dym, C. L. (2000). *Practical optimization: A gentle introduction; fundamentals of modeling and analyzing engineering systems*. New York: Cambridge University Press.

Chvatal, V. (1983). *Linear programming*. Series of Books in the Mathematical Sciences. Paperback, September 15.

Dantzig, G. B. (1998). *Linear programming and extensions*. Cambridge: Ma Princeton University Press.

Eiselt, H. A., & Sandblom, C.-L. (2012). *Operations research: A model-based approach*. Paperback, December 14.

Elmabrouk, O. M. (2012). *Proceedings of the 2012 International Conference on Industrial Engineering and Operations Management Istanbul, Turkey*, July 3–6, 2012.

Gass, S. I. (2010). *Linear programming: Methods and applications* (5th ed.). Dover Books on Computer Science, Paperback, October 21.

Heiman, D. W. (1987) *Operations research as applied to construction*. J:\Documents and Settings \Ian\My Documents\Ianswork\Haidar\C:\Users\cfestin\AppData\Local\Temp\Rar$DI04.808\D. W. Heiman, http://pubsonline.informs.org/doi/abs/10.1287/mantech.1.2.20.

Keil, C. (2008). *A comparison of software packages for verified linear programming*. http://www.ti3.tuhh.de/~keil/pub/ACSPVLP.pdf.

Matousek, J., & Gärtner, B. (2007). *Understanding and using linear programming*. Berlin, Heidelberg: Springer.

Rothlauf, F. (2011). Design of modern heuristics. *Natural Computing Series*. doi:10.1007/978-3-540-72962-4 2. Berlin, Heidelberg: Springer. (Wiley & sons Ltd; June 4, 1998).

Schrijver, A. (1998). *Theory of linear and integer programming*. Wiley Series in Discrete Mathematics and Optimization. Paperback. New York: Wiley, June 4, 1998.

Tam, C. M., Tong, T. K. L., & Zhang, H. (2007). *Decision making and operations research techniques for construction management*. Paperback. Hong Kong: City University of Hong Kong Press, June 30, 2007.

Vanderbei, R. J. (2013). *Linear programming: Foundations and extensions*. Berlin, Heidelberg: Springer.

# Chapter 6
# Techniques for Intelligent Decision Support Systems

**Abstract** Techniques for systems that support intelligent decision making are the new way of applying what is called artificial intelligence, which can assist in solving complex problems in program management. They embody human-like techniques to solve problems ranging from planning, scheduling, and optimization to expert decision making that are difficult to solve using standard mathematical modeling, as described in previous chapters. This chapter will look into knowledge-based systems (also called expert systems) and genetic algorithms which are both widely intelligent systems applied in the construction industry. Knowledge-based systems use computer programming to solve problems associated with human reasoning. They are much simpler than other artificial intelligence methods and can be used effectively in program management where many decisions need to be made, and where the logic can be structured and developed into a software program. Knowledge-based systems are described, together with their applications in the construction industry and especially from the viewpoint of program management. These are empty programs which users can apply to solve their unique problems, thus freeing the hand of the user from the programming. Genetic algorithms, however, are complex techniques that can optimize and find solutions for problems which standard optimization techniques fail to solve. They are search algorithms that mimic the way evolution has progressed by creating a never-ending supply of better generations.

## Introduction

In recent years, a new inter-disciplinary sub-field of interactive computer systems known as artificial intelligence has emerged. Artificial intelligence can be defined as "the field concerned with the computations that connect situations to complex, human-like actions" (Winston 1979). Artificial intelligence encompasses some intelligent behaviors such as deduction, search, learning, and explanation. The main



A.D. Haidar, *Construction Program Management – Decision Making and Optimization Techniques*, DOI 10.1007/978-3-319-20774-2_6

applications of artificial intelligence have been found in programs for processing and understanding natural language, understanding speech, retrieving information, automatic programming, robotics, scene analysis, game playing, fuzzy logic, and proving mathematical theorems. Figure 6.1 shows the different aspects with artificial intelligence.

These applications can be divided into two main subareas:

1. Automated devices, also called robotics, enable the intelligent connection of perception to action. Existing automated systems deal with visual and tactile computer programs to allow automated mechanics to see and manipulate objects in a dynamic environment.
2. Expert systems, also called knowledge-based systems, are automated reasoning systems which attempt to mimic the performance of the reasoning expert.

Further research in artificial intelligence has resulted in other intelligent methods that deal with human biology. These methods are as follows:

1. **Neural networks.** These are algorithms for cognitive problems, such as learning and optimization, which are loosely based on the concepts derived from research into the nature of the brain. Neural computing is applicable to problems whose logical structure is relatively poorly understood.
2. **Genetic algorithms.** These are techniques for solving optimization problems inspired by the theory of evolution and biogenetics. They achieve optimization by emulating the process of natural evolution.

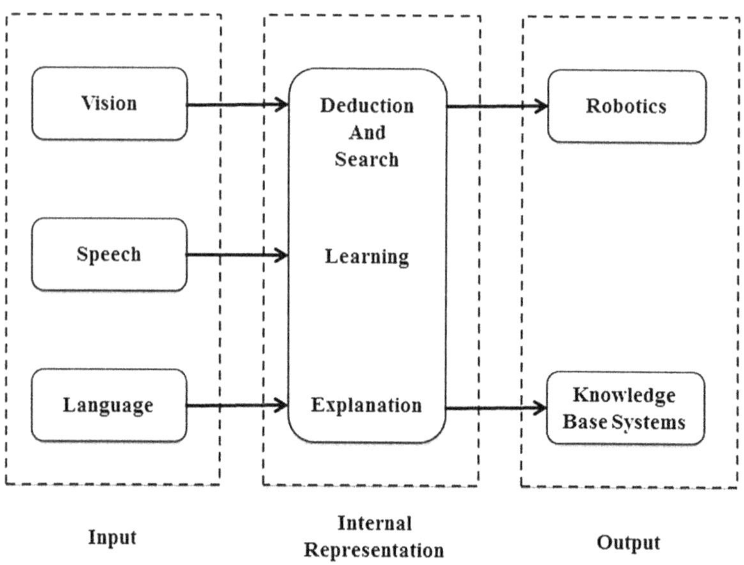


**Fig. 6.1** The different aspects within artificial intelligence

In this chapter, knowledge-based systems and genetic algorithms will be reviewed. Robotics are not considered as their functions are not applicable to processing the behavior of the problems inherent in program management. Neural networks are also not applicable as, at the present state of development, they are inferior to other computing techniques in performing tasks whose structure has been thoroughly analyzed.

## Concept

The concept of the expert system arose in the 1970s when artificial intelligence researchers abandoned or postponed the quest for generally intelligent machines and turned instead to the solution of narrowly focused real-world problems (Yazdani 1986).

Expert systems are interactive computer systems that perform functions normally associated with human intelligence, in order to solve problems typically encountered by human experts in some domain. Expert systems are also called knowledge-based systems because they are based on an extensive body of knowledge about a specific problem area.

Knowledge-based systems use high quality and specialized knowledge in some narrow domain to solve complex problems in that domain. These complex problems lack structured and standard ways of solution and involve substantial use of subjective and experience-based knowledge.

Knowledge-based systems are expanding the applications for computers into new areas of problem solving. These areas involve far more than the numerical data processing of earlier computing systems. While the conventional computer programming approach is still valid in many types of problem and will continue to be so in the future, it is in the application of computers to new problem fields that they are being surpassed by knowledge-based systems (Haidar 1996).

The applications of knowledge-based systems can be of great importance in program management as they encapsulate the knowledge and expertise necessary to solve a great array of problems.

## Definitions

Akerkar and Sajja (2009) stated: "There is no precise definition of knowledge base systems, but they are assumed to be computer systems that can hold human-like knowledge of (in theory) any kind and can process knowledge in a more human-like fashion than do conventional computer systems." Despite this well-defined and observant statement, many authors have provided definitions for knowledge-based systems:

1. They are computer systems that perform functions similar to those normally performed by a human expert
2. They are computer systems that operate by applying an inference mechanism to a body of specialist expertise represented in some knowledge representation formalism
3. They are computer systems that use a representation of human expertise to a particular domain in order to perform functions similar to those normally performed by a human expert in that domain.

Yazdani (1986) defined them as machine systems which embody useful human knowledge in a machine memory in such a way that they can give intelligent advice and also can offer explanations and justifications on demand.

In a somewhat authoritative attempt at as definition of knowledge-based systems, Frost (1986) suggested that they are means of capturing the knowledge of experts in the form of programs and data where disagreements among the experts are settled by mediation and the results refined so as to extract the essence of their knowledge in such a way that it can be used by less experienced people within the field.

These definitions were summarized by Haidar (1996) as "intelligent computer programs that use knowledge and inference procedures to solve problems that are difficult enough to require significant human expertise for their solution. Knowledge necessary to perform at such a level, plus the inference procedures used, can be thought of as a model of the expertise of the best practitioners in the field."

## Architecture and Components

In general, a knowledge-based system is a computer program that:

1. contains human knowledge
2. is able to give advice by inferring from this knowledge
3. can justify the advice given
4. in which the knowledge can be maintained independently of the program.

Figure 6.2 shows the diagram of typical knowledge-based system with its main components.

## Knowledge Base

Knowledge base comprises the store of knowledge and expertise that has been programmed into the system and which is relevant to the problem under investigation. In other words, it is essentially an empty framework, which becomes a working knowledge base with the addition of the expert's know-how. The main

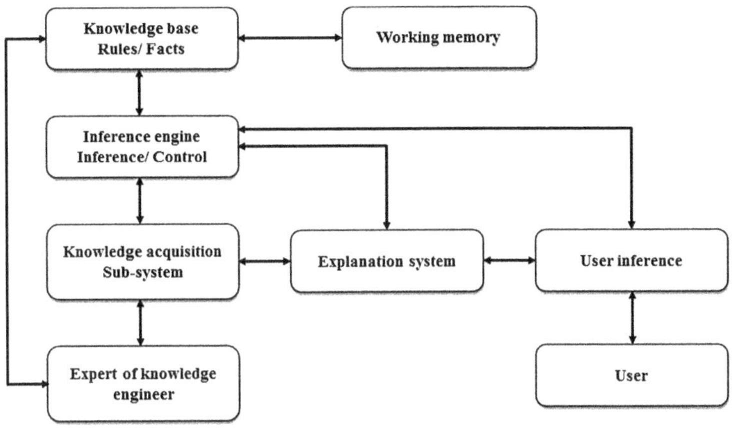


**Fig. 6.2** Diagram of a typical knowledge-based system

difficulty in developing a knowledge-based system is in representing the knowledge. The knowledge representation problem concerns the decision and on how to encode knowledge so that the computer can use it. In the early stages, knowledge-based systems used rule-based formalism, in which knowledge can be expressed in the form of conditions or action rules.

Since the development of rule-based systems, other methods of knowledge representation have evolved:

(a) *Semantic nets*—these represent relations among objects in the domain by providing links between nodes
(b) *Frames*—these are generalized record structures which may have default values and may have actions coded as the values of certain fields or slots
(c) *Horn clauses*—these are a form of predicate logic on which Prolog computer language is based and, using the Prolog system, can perform inferences.

The main difference between rule-based programming and normal programming is the declarative nature of the former. While conventional programming is procedural by nature, in that the developer has to specify the order of the program flow, rule-based programming allows rules to be declared and maintained without specifying when and how to apply them. To achieve this, rule-based programming has three distinct components:

1. a rule base
2. an inference engine
3. a user interface.

These components are expanded upon below.

## The Rule Base

Knowledge (particularly surface knowledge) can be expressed in the form of "Situation => Action" rules, where a situation is a set of conditions (also known as production rules). The syntax of such rules is as follows:

IF Condition1 and Condition2 and … THEN Action1 and Action2 and…

Examples of the rule-based syntax:

- IF: large concrete pour is scheduled THEN: check the capacity of batch plants
- IF: activity has no float time (TF = 0) THEN: activity is on the critical path
- IF: delivery of material is scheduled to be late THEN: activity will be delayed.

Each condition (1, 2, … etc.) can be true or false and represents a check on the value of a parameter or attribute. Variables appearing in conditions can be of type integer, real, string or date, and the operators linking the variables to the values can be symbolic, such as "is" and "less than."

A condition is normally "AND" with other conditions as shown in the syntax above. In addition, as in some development tools, a condition can consist of "OR" sub-conditions. Each condition can only have two possible logical values: True or False.

The Action part of the rule can contain assignments of further conditions, calculations, report messages, etc. The rule syntax is sometimes extended to include an ELSE part following the THEN part.

A rule base or a knowledge base is a list of the condition rules declaring the know-how in a particular domain (area of interest). The declarative nature of the rule base enables the user to define the rules without stating how and when they can be used. The application of rules is controlled by an inference engine and control rules, as described in the next section. The often listed advantages of rule-based programming over conventional programming are the simplicity of syntax, understandability, and ease of maintenance, since rules can be maintained independently of each other.

## Knowledge Acquisition

Knowledge acquisition has been defined as "the transfer and transformation of potential problem-solving expertise from some knowledge source into a problem-solving expertise."

Knowledge acquisition is the process by which expert system developers find the knowledge that domain experts use to perform the task of interest. This knowledge is then implemented to form an expert system. The essential part of an expert system is its knowledge, and therefore, knowledge acquisition is probably the most important task in the development of an expert system. The knowledge acquisition process usually involves the following stages:

(a) Identify the knowledge domain,
(b) Examine the proposed system goals,
(c) Locate the sources of domain knowledge,
(d) Define domain boundaries,
(e) Elicit the knowledge, and
(f) Review and analyze the acquired knowledge.

Knowledge induction and machine learning techniques have been developed to automatically gather rules from experts to assist in obtaining their knowledge. Induction tools allow a knowledge-based system to induce its own knowledge from a set of data or examples.

## Working Memory

The working memory stores the information provided by the user during a consultation, along with the conclusions, sub-conclusions, and reasoning that the system is using at any particular time. The working memory can usually be executed easily to provide the user with extra information concerning the system operation.

## Inference Engine

The inference engine's main function is to compare the stored facts with the current conditions in order to select an appropriate action or inference. It performs the task of inferring the required decisions (conclusions) from a rule base. To understand the operation of the inference engine, we will consider a rule base relating to dressing in cold weather.

RULE1
IF weather is cold AND you are going out THEN wear a coat.
RULE2
IF temp <50 °C THEN weather is cold.
RULE3
IF it is raining THEN carry an umbrella.

There are two possible inference strategies used by an inference engine to infer (derive decisions) from the above rule base: forward chaining inference and backward chaining inference.

Forward chaining is also referred to as "data driven" and is the simpler of the two strategies. With a forward chaining strategy, the inference engine scans the rule base searching for rules that can fire (conditions are satisfied) given any initially available data. Since firing rules can generate further data, when the conclusion of a fired rule is a condition of another rule, the forward chaining process is repeated until no further rules can be fired. Hence, for the above rule base, given that "it is raining," the only conclusion we can infer by forward chaining (RULE3) is "carry an umbrella." On the other hand, given that the temperature is below five degrees, then forward chaining infers that "weather is cold." To address situations where there is no initial data supplied, "ask rules" are often added to the rule base to request data from the user.

Backward chaining is also referred to as "goal driven" inference. The inference strategy starts from a goal (a conclusion) and works backward to prove whether the goal is true or false by evaluating all conditions needed to satisfy the required goal. For the above rule base, if we want to investigate the goal "wear a coat" (i.e., should we wear a coat or not?), then the inference engine will commence the search for a rule with that conclusion, finding RULE1. In order to verify whether this conclusion is true or false, we need to verify the conditions of RULE1. The first condition is "weather is cold." The inference engine will now treat this condition as its goal and will search for a rule with this conclusion, finding RULE2. The inference engine will now have to verify the condition of RULE2; "temp <5 °C." Since there are no rules with such a conclusion, the inference engine will call the user interface to obtain a value for temperature from the user. Assuming that the user confirms that temperature is less than five degrees, then the conclusion of RULE2 is true (i.e., weather is cold). The inference engine will return to RULE1, having proved that the first condition is true, to verify the second condition "you are going out." Since there are no rules with this conclusion, the inference engine will once again call the user interface for a value, assuming this condition is true then the conclusion of RULE1 is true (i.e., we have to wear a coat).

The inference engines of commercially available rule-based development systems support one or both inference strategies. In systems supporting both forward and backward chaining strategies, a certain degree of control over the order of inference can be exercised by the developer. This is achieved by "agendas" or "control rules" to control the flow of inference and by optionally disabling forward firing of certain rules. An additional feature to control inference is that of using "demons" or "when rules." These are rules that fire when their conditions are true regardless of where in the inference cycle the engine is. These inference control options allow the developers of knowledge-based systems to impose a hierarchical knowledge control model on rule bases.

## User Interface

The user interface is used to supply the inference engine with information on conditions for which no rules can be found. It is used to communicate any messages, reports, or conclusions to the user. The use of natural language for input and output is now widespread in most user interfaces. An in-built explanation model allows the system to justify its conclusions upon request.

## Features

The features of knowledge-based systems that distinguish them from conventional computer programs can be summarized as follows:

1. They know a great deal about a limited but useful area of interest
2. They give advice conversationally in the manner of a consultant and can understand and respond to simple questions posed in plain (though perhaps specialized) English language
3. Their knowledge is embodied as separate modules containing sets of rules with corresponding actions. This feature makes it easier for correction of deficiencies or errors in their knowledge bases as well as the acquisition of new knowledge
4. The questions posed by the system are limited to those that are relevant to a particular line of reasoning. Thus, if at any time the systems decide they have sufficient information to arrive at a conclusion, they stop asking questions
5. Above all, knowledge-based systems can credibly explain and justify their reasoning.

## Limitations

The limitations of applying knowledge-based systems are that they cannot deal effectively with problems that require none of their advantages. These are as follows:

1. Problems beyond the limits of their knowledge
2. Problems that are of a well-structured numerical in nature
3. Problems that are too simple
4. Problems that are too complex
5. Problems with inadequate knowledge
6. Problems with the "pregnant male syndrome" where the expert system deduces the patient is pregnant because of laboratory results but has never been informed that only females become pregnant and so erroneously deduces that a male patient is pregnant
7. Problems in "wide and shallow" domains.

## System Environments

A number of different approaches and programming may be used to develop a knowledge-based system. They include the following.

(a) *Standard programming languages* such as Pascal, C, Fortran, and BASIC. They are high-level programming languages with their own compiler or interpreter and a runtime environment for writing and debugging programs. These languages are general and offer a certain amount of flexibility in developing knowledge representation and inference schemes. The major disadvantage of developing a system using a language is the time and effort involved in programming the system from scratch. Only a few individuals have insisted on applying conventional languages to tasks for which they were not designed. Almost all development of knowledge-based systems has taken place using other development media.
(b) *Artificial intelligence languages* such as Lisp and Prolog. They are more suited to the task than conventional languages, but still require programming expertise, especially on facilities such as the user interface for which the language may not be particularly well suited.
(c) *Development environments* such as OPS83, RLL, Rosie, Keystone, Knowledge Craft, Art, and KEE. These are one level higher than the above computer languages. They are usually rule-based programming languages and offer a variety of methods for representation and control of the reasoning process. Unfortunately, these are generally large items of software which often require specialized hardware and are correspondingly expensive.
(d) *Knowledge-based systems shells.* These are skeletal systems containing the knowledge representation and inference schemes. A shell can be thought of as a knowledge-based system with all the domain-specific knowledge removed and a facility for entering a new knowledge base provided. In other words, it is an application-building tool which provides a software environment for expressing knowledge and an inference engine to interpret and apply this knowledge to a particular problem. By insulating the users from the internal representation and logic, shells permit development of systems by people who are not conversant with programming.

The expert system shell is responsible for the following:

- managing and processing input and service requests from users, and generating output
- supporting the creation and modification of inference rules by knowledge engineers
- translating the inference rules created into machine-readable forms
- processing the information given by the user and the application layer modules, and relating such information to the concepts contained in the knowledge base through the inference rules
- providing solutions for particular problems within the scope of its integrated knowledge domain
- providing facilities for uncertain reasoning
- providing facilities for knowledge representation (the knowledge representation knowledge) and editing the content of the knowledge base (knowledge-based editor)

- providing low-level support to expert system components (such as retrieving metadata from and saving metadata to the knowledge base, building abstract syntax trees during the translation of inference rules, and other similar functions).

In essence, the system shell is indifferent to the rules it executes. This distinction is very important, because it means that the expert system shell can be applied to many different problem domains with little or no change. It also means that editing (adding or modifying) the rules of an expert system can implement some changes in the way the program behaves, without necessarily affecting the underlying controlling component—the shell. As was observed, expert system shells provide methods of building expert systems without extensive knowledge of programming through mechanisms that:

- input the decisions, questions, and rules that are followed
- structure or develop a knowledge database that can be manipulated by subsequent parts of the system
- verify possible violations of surface validity
- operate the "inference engine" that operates on the rules, poses the questions to the users, and determines whether a particular decision is valid.

Some of the knowledge-based systems currently available are EMYCIN (Empty MYCIN), ESIE (Expert System Inference Engine), Savoir, Xi-Plus, Leonardo, XpertRule, Exsys, Guru, RuleMaster, Twaice, Crysta, EspAdvospr, and Sage.

The difference between the knowledge-based shell and the final knowledge base is shown in Fig. 6.3.

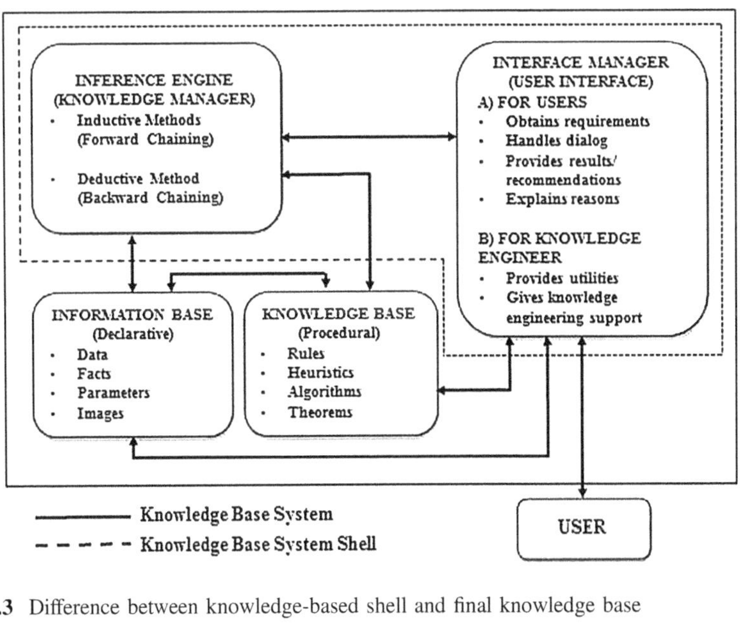


**Fig. 6.3** Difference between knowledge-based shell and final knowledge base

## Knowledge Management in Construction

It was in the mid-1980s that people began to appreciate the increasingly prominent role of knowledge in the competitive environment, with the emergence of knowledge base as a distinctive management method. This new approach recognizes knowledge as one of the most valuable assets of an organization in contrast to the traditional economic view, which recognizes knowledge as something external to the company and with no connection to the economic process. Moreover, this approach gives a clear structure for managing knowledge, with greater emphasis on the knowledge itself and with a hierarchy above information and data. Thus, it is possible to define knowledge management as the way in which organizations create, capture, and use knowledge to achieve their objectives.

Construction companies obtain most of their knowledge from the projects they undertake. However, the knowledge generated within each project is finally stored in reports that remarkably few read, or is lost because the people involved move to a new project, leave the company or retire, taking with them not only their tacit knowledge but also a potential source of competitive advantage (Arinze and Partovi 1992).

Regarding how to manage knowledge in the selection of construction methods, construction companies use the knowledge of individuals to carry out this process. There is not an organization-based learning process that allows acquisition of the relevant knowledge.

Knowledge management in the construction industry is a focus of different types of research work, for example, studies that have tried to understand how to implement knowledge management in construction companies, and also the perceptions of people about this topic. Learning has also undergone some studies as well as the development of knowledge management models, the development of systems to store and share knowledge, and the development of knowledge maps (Alfredo et al. 1990).

Other lines of research have focused on understanding the impact of technology in data capture in the field, in the management of documentation, and in the development of methodologies for the capture and reuse of the knowledge created in projects. Other researchers have studied how to share tacit knowledge within areas of practice and how to make a live capture and reuse of project knowledge. The importance of collaborative knowledge management has also been addressed, and in recent years there have been studies about the use of mobile technologies in construction, among others.

## Knowledge-Based Systems in Program Management

Knowledge-based systems can be developed as a tool for program managers. Expert system shells can allow knowledge-based systems to be developed and used on construction sites effectively and with minimum programming.

Construction experts can provide knowledge which can define the best crew size, the expected productivity, and the effects of production bottlenecks. This information can be incorporated in a knowledge base and used as an aid in determining when production problems are occurring in the field. The information can also be used in the pre-construction period by providing managers and estimators with expert knowledge of expected productivity and required crew size.

Construction companies have difficulties in managing the information and knowledge associated with construction programs, combined with the fact that much of the information about previous projects is not reused because there are not adequate mechanisms for its storage. In addition, the knowledge created in the field is not usually shared, which tends to result in its loss. This situation eventually affects decision-making processes because correct decisions are the result of the careful management and analysis of the information and knowledge available.

In the construction industry, the program team members gain experience and knowledge in their specific functional area. The personnel, over a period of time, through various construction projects, develop a huge expert knowledge base from the practices and it is well established that having access to a broad knowledge base is often essential before a construction problem can be understood and resolved. When experienced construction personnel leave the company, so also does their expert knowledge base. Most of the highly relevant information about the construction process typically remains locked within the person (Ferrada et al. 2013).

Construction methods are the means used to transform resources into constructed product. Programming and management techniques are of little value for a program if construction methods are not the most optimal in terms of cost or are not safe to run. The selection of construction methods affects not only the selection of the activities and their work sequence, but also the duration of the program. In construction, this process is highly iterative and requires the construction team to examine a variety of data sources as well as to tap into its own experience base to formulate a set of efficient methods.

In cases like this, when a decision problem has at least two conflicting criteria and at least two solution criteria, the problem is considered a multi-criteria decision analysis. Given the impact construction methods have on productivity, quality and cost, their selection is a key decision for the proper development of a construction program, and it is one of the main factors affecting the productivity and efficiency of the program. Also, it is considered as one of potential areas of productivity loss.

These facts highlight the significance of an appropriate selection of construction methods for a program since deficient methods for executing the work can cause significant losses of productivity on the different projects within a program. It is a powerful tool that can help this industry to innovate and improve its performance. The appropriate selection of construction methods to be used during the execution of a construction program is a major determinant of high productivity, but sometimes this selection process is performed without the care and the systematic approach that it deserves, bringing negative consequences. Therefore, program managers should propose a knowledge management approach that will enable the intelligent use of corporate experience and information and help to improve the selection of construction methods for the program.

## Applications

One of the first knowledge-based systems was MYCIN, developed in 1974, in which subjective and heuristic knowledge of expert physicians was used to diagnose infectious diseases and provide antimicrobial therapy. Since the development of MYCIN, knowledge-based systems are under active development in a wide range of disciplines related to science and engineering. Some major applications of knowledge-based systems related to the construction industry generally and to project management specifically are as follows:

(a) Muñoz-Avila et al. (1987) developed an intelligent planning system to assist project planners in the creation of work breakdown structures that could significantly expedite the planning process and increase its chances of success. Their proposal is based on two areas of research: (1) automated hierarchical planning systems, and (2) knowledge management (KM), which advocates reusing previous problem-solving and decision-making experiences to improve organizational processes. Building on this foundation, they present architecture for knowledge-based project planning. The system architecture employs an integrated set of methodologies, including hierarchical plan generation and case retrieval, for reusing experience to support a project planner in the creation of a work breakdown structure (WBS). Their study is organized into four components or sections. Section 1 examines hierarchical plan representation and presents associated plan generation techniques. Section 2 then compares the WBS and hierarchical plan representations. Section 3 presents architecture for a knowledge-based project planning system with the ability to support automated plan generation. Finally, Sect. 4 describes a methodology for developing a knowledge base for such a system.

(b) Alfredo et al. (1990) present a knowledge management approach that includes both a knowledge-based application framework and a prototype system developed to verify the framework of the knowledge management approach. The objective of the system is to support decision making for the correct selection of construction methods for a construction program. The study presents, as well, the main background on the selection of construction methods and knowledge management, the conceptual development and main features of the proposed knowledge system and, finally, the operation of the prototype system used to validate the proposal.

(c) One of the major roles undertaken by a project manager is the management of the risk of a project. However, this duty is particularly complex and inefficient if good risk management has not been done from the beginning of the project.

Alfredo et al. (1990) proposed an effective and efficient risk management approach that includes a proper and systematic methodology and, more importantly, knowledge, and experience. Their study addresses the problems of risk management in construction projects using a knowledge-based approach and proposes a methodology based on a threefold arrangement that includes the modeling of the risk management function, its evaluation, and the availability of a best practices model. A major preliminary conclusion of their research is the fact that risk management in construction projects is still very ineffective and that the main cause of this situation is the lack of knowledge. It is expected that the application of the proposed approach will allow clients and contractors to develop a project's risk management function based on best practices and also to improve the performance of this function.

(d) Arinze and Partovi (1992) propose the development of a knowledge-based system to enable the incorporation of expertise into project management. Their proposed system represents an alternative to traditional algorithmic approaches. It enables the user to selectively modify and constrain single activities or the entire network by specified amounts, reconfigure the network, and computer resource usage in each case. The system, therefore, represents a potentially useful decision support tool for performing exploratory and sensitivity analyses in managing large projects and a training tool for less experienced planners.

(e) Arain and Pheng (2006) in their study, describe the framework for developing a knowledge-based decision support system (KBDSS) for making more informed decisions for managing variation orders in institutional buildings. The KBDSS framework consists of two main components, i.e., a knowledge base and a decision support shell. The database was developed through collecting data from source documents of 80 institutional projects, a questionnaire survey, a literature review, and in-depth interview sessions with the professionals who were involved in these institutional projects. The KBDSS developed is capable of displaying variations and their relevant details, a variety of filtered knowledge, and various analyses of available knowledge. This approach would eventually lead the decision maker to the suggested controls for variations and assist in selecting the most appropriate controls. The KBDSS developed by Arain and Pheng would assist project managers to:

- provide accurate and timely information for decision making
- provide a user-friendly system for analyzing and selecting the controls for variation orders for institutional buildings
- assist building professionals in developing an effective variation management system
- help them to take proactive measures for reducing variation orders.

A knowledge-based system to support this decision-making process is proposed to:

1. Define and design the system. Semi-structured interviews need to be conducted within different construction companies with the purpose of studying the way that the method selection process is carried out in practice, and the knowledge associated with it.
2. Develop a prototype of a construction method's knowledge system and then validate it with construction industry professionals.

## Genetic Algorithms

### *Concept*

Like knowledge-based systems, genetic algorithms have emerged from laboratories in the 1970s and are gaining popularity as strong tools for optimization problems too difficult to solve using conventional methods. These algorithms are excellent at exploring large search spaces for optimal or near optimal solutions.

Genetic algorithms are search algorithms based on the mechanics of natural selection and natural genetics. They combine survival of the fittest among string structures with a structured yet randomized information exchange to form a search algorithm with some of the innovative flair of human search. In every generation produced by the genetic algorithms structure, a new set of artificial creatures (strings) is created using bits and pieces of the fittest of the old. An occasional new part is tried to make sure fit parts have not been missed out in the process. When randomized, genetic algorithms are a complex search optimization procedure as they efficiently exploit past information on new search points with expected improved performance.

Genetic algorithms are different from conventional optimization and search procedures in four ways:

1. They work with a coding of the parameter set, not the parameters themselves. Genetic algorithms manipulate decision or control variable representations at the string level to exploit similarities among high-performance strings, and therefore, they are difficult to fool even when the function is difficult for traditional methods.
2. They search from a population, not a single point. In this way, by maintaining a population of well-adapted sample points, genetic algorithms reduce the probability of reaching a false peak.
3. They use payoff (objective function) information, not derivatives, or other auxiliary knowledge. Genetic algorithms achieve much of their breadth by ignoring information except that concerning payoff. Therefore, they remain general by exploiting information available in any search problem, even where the necessary information is not available or is difficult to obtain. Genetic algorithms process similarities in the underlying coding together with

information ranking the structures according to their survival capability in the current environment. By exploiting such widely available information, genetic algorithms may be applied in virtually any problem.

4. They use probabilistic operators, not deterministic rules. Genetic algorithms use random choice to guide a highly exploitative search through its randomized stochastic operators.

Genetic algorithms can converge quickly on solutions in large search spaces through their remarkable ability to focus their attention on the most promising parts of a solution space and their ability to combine strings containing partial solutions.

Genetic algorithms have demonstrated their ability to make breakthroughs in the design of many complex systems in various fields as they make it possible to explore a far greater range of potential solutions to a problem than do conventional programs.

## Process

A potential solution to the genetic algorithms optimization problem is represented as a set of binary values. These values are analogous to the individual genes. A chromosome represents a sequence of genes which require optimization.

Optimization of genetic algorithms begins with an initial generation. Each chromosome is assigned a fitness based on its ability to meet the objectives and constraints of the problem. The fittest chromosomes are more likely to be selected to pass their genes to the next generation. The generation undergoes a series of random processes on each iteration of the genetic algorithm and forms a new generation. In Fig. 6.4, A, B,..., N represent the chromosomes of the initial

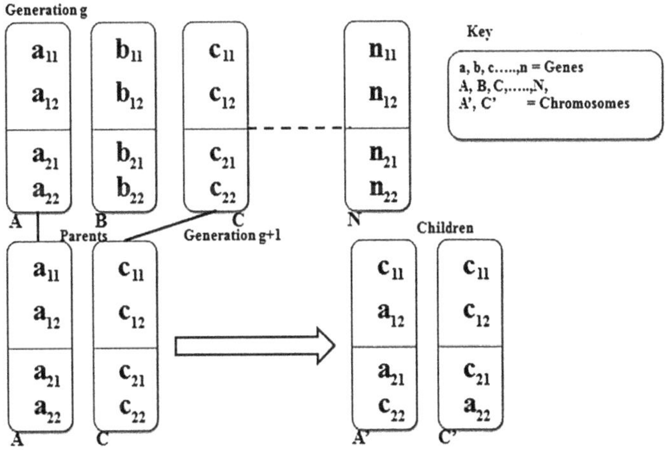


**Fig. 6.4** The process of genetic algorithms

generation. a, b, c… n represent the genes that form the chromosomes and the binary numbers selected randomly within the genetic algorithms engine. A′ and C′ represent the chromosomes of the next generation formed by mixing the genes of the fittest chromosomes in the initial generation.

The two basic steps in developing solutions using genetic algorithms are an appropriate representation of the problem and a method of assessing the effectiveness of a solution.

Each parameter in the genetic algorithms formulation problem is represented by a gene which can be constrained by a minimum and a maximum value. A cost function is defined which derives a value for the cost of the solution from a given set of genes. The cost function, in this instance, represents the objective function that needs minimizing or maximizing in the optimization problem. Penalties or large values are added to lethal constraints which will give the chromosomes that violate their constraints a very high cost and will be overlooked in the parent selection process. The function of the genetic algorithms would be to find a set of genes that would minimize or maximize the defined cost parameter.

## Evolution Operators

The ability of genetic algorithms to focus their attention on the most promising parts of a solution space is a direct outcome of their ability to combine strings containing partial solutions (Haidar 1996). After the initial generation, the random search of better solutions for the rest of the generations is controlled through the evolution operators. There are three main operators, namely, crossover, mutation, and adaptation.

1. *Crossover* is the random recombination of the genes of two parents to form a *child*. In a genetic algorithm problem, crossover is a random binary combination of the genes of two separate chromosomes to provide a new "*child*" chromosome. In this process, the selection of the "*parent*" chromosomes is biased to the more effective (fit) chromosomes.
   In this process, a random point(s) along the strings of two genes is selected at random and the portions to one side of that point are exchanged between the genes to create a new gene. Figure 6.5 shows the rearrangement of the genes of two chromosomes.
2. *Mutation* is used to add new genetic materials to the gene pool and is also part of the mechanism of retaining bad values by creating a whole new number in the chromosome. Mutation alone does not generally advance the search for a solution, but it does provide insurance against the development of a uniform population incapable of further evolution. It is the action of random mutation that lets genetic algorithms avoid being captured by local minima.

| 1 | 0 | 1 | 0 | 1 | 0 | 1 | 0 | 0 | 1 | 1 | 0 | 1 | 0 | 1 |
|---|---|---|---|---|---|---|---|---|---|---|---|---|---|---|

Chromosomes 1

| 1 | 1 | 0 | 0 | 0 | 1 | 0 | 1 | 1 | 0 | 1 | 1 | 1 | 0 | 0 |
|---|---|---|---|---|---|---|---|---|---|---|---|---|---|---|

Chromosomes 2

| 1 | 0 | 1 | 0 | 0 | 1 | 0 | 1 | 1 | 0 | 1 | 0 | 1 | 0 | 1 |
|---|---|---|---|---|---|---|---|---|---|---|---|---|---|---|

Child Chromosomes

**Fig. 6.5** Crossover between two chromosomes

3. *Adaptation*, like mutation, is a random change to the value or order of genes within the chromosomes. However, it is different from mutation as it retains only improved values. As such, adaptation is a wise mutation which helps to accelerate the search for the solution.

## Features

The main features of genetic algorithms are as follows:

1. Genetic algorithms provide an efficient optimization mechanism for problems characterized by many constraints, uncertainty, and an abundance of feasible solutions.
2. A properly designed genetic algorithms program works fast since the effect of successive evolution operators of two acceptable solutions quickly coalesces the best features, resulting in a better solution.
3. Problem-specific knowledge, if available, can be incorporated to guide the search through the representation and the operators.
4. Genetic algorithms can be applied to solve poorly understood, loosely defined problems, which are beyond the scope of most of the traditional search methods.

## Limitations

The main limitations of genetic algorithms are as follows:

1. The complexity and the amount of time needed for programming.
2. They are good for finding a "good enough" near-best solution very quickly, but do not guarantee a single best solution. However, this can be achieved by trying the system numerous times and observing the performance of the solutions generated.

3. The size of the generation can be critical in many applications of genetic algorithms. A small generation may force the genetic algorithms to converge too quickly without achieving the optimal solution. If the generation is too large, the waiting time for an improvement might be too long (Haidar 1996).

## Applications

Genetic algorithms have been practically applied and tested in a wide variety of engineering contexts.

### *Multiple Constraint Scheduling*

Sriprasert and Dawood (2003) developed a genetic algorithms to solve multi-constraint optimization problems. Their study introduces a methodology termed “multi-constraint scheduling” in which four major groups of construction constraints are considered and it comprises the following:

(1) Contract constraints—time, cost, quality, special agreements
(2) Physical constraints—technological dependency, space, safety, environment
(3) Resource constraints—availability, capacity, perfection, continuity
(4) Information constraints—availability and perfection.

Given multiple constraints such as activity dependency, limited working area, and resource and information readiness, the genetic algorithms alters tasks’ priorities and construction methods so as to arrive at an optimum or near-optimum set of project duration, cost, and smooth resource profiles. Several experiments done by Sriprasert and Dawood confirmed that the genetic algorithms can provide near-optimum solutions within acceptable searching time. Possible improvements to this research are further suggested in their study.

In practice, basic PERT and CPM scheduling techniques have proven to be only helpful when the project deadline is not fixed and the resources are not constrained by either availability or time. To deal with project resources, Sriprasert and Dawood used two main types of techniques, namely, resource allocation and resource leveling. Resource allocation (sometimes referred to as constrained-resource scheduling) attempts to reschedule the project tasks so that a limited number of resources can be efficiently utilized while keeping the unavoidable extension of the project to a minimum. Resource leveling (often referred to as resource smoothing), on the other hand, attempts to reduce the sharp variations among the peaks and valleys in the resource demand histogram while maintaining the original project duration. For each of these two problems, there are many heuristic rules that are simple, manageable for practical-size projects, and utilized by almost all commercial planning and scheduling software.

Despite these benefits, however, heuristic rules perform with varying effectiveness when used on different networks and by no means guarantee an optimum. The formulation of a multi-constraint optimization problem using GA and the practical development of a GA-based application in the MS project have also been described by Sriprasert and Dawood. A case example and the advantages of the developed GA over other methods have been presented. It is envisaged that successful implementation of the overall applications for this particular complex problem will assist project planners to produce more reliable plans which will, in turn, promote effective co-ordination across supply chains and various trades at the construction work face. Despite these benefits, certain aspects that need further research and development are as follows:

1. Consideration of more constraints such as safety and environment;
2. Investigation of various formulation techniques for the multi-objective optimization problem; and
3. Implementation of advanced GA mechanisms for n-points crossover and mutation to cope with the problem of larger complex projects.

## Time–Cost Trade-off

Azaron et al. (2005) developed a multi-objective model for the time–cost trade-off problem in PERT networks with generalized Erlang distributions of activity durations, using a genetic algorithm application.

Time–cost trade-off analysis is one of the most important aspects of construction project planning and control. Given a construction project network, the objective is to select appropriate resources and methods so that the tasks of a project can be completed within the required duration and minimum cost. In general, the less expensive the resources used, the longer it takes to complete an activity. In this study, the time–cost trade-off problem is modeled using a set of construction method options. Deriving from historical data, several possible options with different sets of activity durations, resource requirements, and direct resource costs are assigned to each activity in the project network. The genetic algorithms randomly search through possible combinations of options assigned to each activity and evaluate the fitness of time and cost on the basis of the weights and criteria presented.

The mean duration of each activity is assumed to be a non-increasing function and the direct cost of each activity is assumed to be a non-decreasing function of the amount of resource allocated to it. The decision variables of the model are the allocated resource quantities. The problem is formulated as a multi-objective optimal control problem that involves four conflicting objective functions. These objective functions are the project direct cost (to be minimized), the mean of the project completion time (min.), the variance of the project completion time (min.) and the probability that the project completion time does not exceed a certain threshold (max.).

It is impossible to solve this problem optimally. Therefore, Azaron applied a "Genetic Algorithm for Numerical Optimizations of Constrained Problems" (GENOCOP) to solve this multi-objective problem using a goal attainment technique. Several factorial experiments are performed to identify appropriate genetic algorithm parameters that produce the best results within a given execution time in the three typical cases with different configurations. Finally, Azaron compared the genetic algorithms results against the results of a discrete-time approximation method for solving the original optimal control problem.

## Construction Material Delivery Schedules

Georgy and Basily (2008) developed a systematic procedure and a computerized tool for optimizing the delivery and inventory of materials, as part of a comprehensive material management system in construction projects. The study partially fulfills a long-sought research need for developing comprehensive material management systems specifically tailored to construction projects. The system takes into account several parameters that are not typically incorporated in the economic order quantity models for material management. Furthermore, practicality of the introduced system is augmented by the fact that it is interlinked with one of the most commonly used scheduling softwares.

A newly devised approach that employs genetic algorithms for the optimization of material delivery schedules and their associated inventory control is also presented by Georgy and Basily. The approach is based on the project material requirement plans and employs an objective function that minimizes the total costs associated with material deliveries. Furthermore, the computer system developed by Georgy and Basily is used to examine and validate the adopted approach.

Genetic algorithms proved to be a satisfactory approach for optimizing material delivery schedules and their associated inventory levels. The selected case study particularly showed the system to produce material delivery plans that have reduced costs compared with their actual counterparts. Also, the computer processing time for developing the optimized plans was fairly minimal, which promotes its practical use.

## Resource Leveling

To meet the physical limits of construction resources, to avoid day-to-day fluctuation in resource demands and to maintain an even flow of application for construction resources, resource leveling is applied in the construction industry. Traditional resource-leveling models assume activity durations to be deterministic. Nevertheless, activity duration may be uncertain, owing to variations in the overall environment such as weather, site congestion, and productivity level.

An optimal construction resource-leveling model is needed, in which the combinative effects of both uncertain activity duration and resource leveling are taken into consideration. A simulation is used to model the uncertainties of activity duration. A searching technique using genetic algorithms is then adopted to search for the impact of uncertain activity durations on the probabilistic optimal resource-leveling indices. The model can effectively provide probabilistic optimal resource-leveling indices for multiple construction resources subjected to the objective of resource leveling and the impact of influence factors on the probabilistic resource-leveling scheduling problems (Fathi and Afshar 2008; Hegazy 1999; Iranagh and Sonmez 2012).

## Finance-Based Scheduling

The capability to obtain sufficient cash at the right time is considered one of the most common and critical challenges that a contractor usually faces during the execution of any construction program. As a result, cash must be thought of as a limited resource because of the importance of its procurement for contractors. During the period of any program, a contractor might not carry out any work that has no cash availability despite the commitment to schedules. This clear principle of operation makes the establishment of a balance between financing needs and available cash throughout the program duration a very vital concept in producing realistic schedules. In other words, if sufficient cash is not available to the contractor, the program duration will be increased, leading to an increase in overheads and a decrease in the contractor's profit.

Therefore, a sound and well-managed program for cash flow scheduling should be established in order to allow the contractor to identify his/her cash needs during each period of the constructed project(s).

However, due to the distinctive feature of cash, none of the previous studies can be used to devise cash-constrained schedules. The distinctive feature is that while cash, like any other resource, is being used to carry out construction works, the completed construction works generate the same resource of cash which is used to finance the remaining activities of the project. As a result, some research efforts have integrated critical path method schedules with cash flow models to devise what is called "finance-based scheduling."

In order to achieve this main objective; sub-objectives are to be attained as follows:

1. Construct a CPM scheduling model and the associated cash flow.
2. Develop an evolutionary optimization algorithm for solving scheduling problems.
3. Implement and test the integrated scheduling multi-objective algorithm.

Lack of financing and cash deficit are considered as a primary threat to the contractor's financial management. Therefore, in case of insufficient cash, many contractors find it difficult to stick with the project schedule, thus leading to extra overhead costs and liquidated damages. Contractors mainly deal with the project scheduling and financing as two independent functions of construction project management.

El-Abbasy et al. (2012) developed a multi-objective elitist genetic algorithms for solving finance-based scheduling problems of multi-projects with multi-mode activities. A Critical Path Method scheduling model is constructed with its associated cash flow to calculate the values of the multiple objectives. The problem involves the minimization of conflicting objectives: duration of multiple projects, financing costs, and maximum negative cumulative balance. The designed optimization model performs operations of the genetic algorithm in three main phases:

1. population initialization,
2. fitness evaluation, and
3. generation evolution.

The model can be considered relevant for practitioners to use in large construction programs to make decisions regarding financing.

The methodology followed in this study in order to achieve the main objectives can be summarized as follows:

1. Build a Critical Path Method scheduling model and its associated cash flow in order to calculate the values of the multiple objectives to be optimized.
2. Develop a genetic algorithms-based optimization model to optimize the objectives without violating the set constraint. The model to be designed will perform genetic algorithm operations in three main phases: (1) population initialization, (2) fitness evaluation, and (3) generation evolution.
3. Implement the model using a computer language program to perform the optimization process.

## References

Akerkar, R., & Sajja, P. (2009). *Knowledge-based systems*. Canada: Jones & Bartlett Learning.

Alfredo, F. S., Ferrada, X., Howard, R., & Rubio, L. (1990). *Risk management in construction projects: A knowledge-based approach*. Amsterdam: Elsevier. http://www.sciencedirect.com/science/article/pii/S1877042814021648.

Arain, F. M., & Pheng, L. S. (2006). A framework for developing a knowledge-based decision support system for management of variation orders for institutional buildings. *Special Issue Decision Support Systems for Infrastructure Management* (Vol. 11, pp. 285–310). http://www.itcon.org/2006/21.

Arinze, B., & Partovi, F. Y. (1992). A knowledge-based decision support system for project management Bay. *Journal Computer and Operation Research, 19*(5), 321–334.

Azaron, A., Perkgoz C., & Sakawa M. (2005). A genetic algorithm approach for the time-cost trade-off in PERT networks. *Applied mathematics and computation* (Vol. 168(2), pp. 1317–1339), September 15, 2005.

El-Abbasy, M., Zayed, T., & Elazouni, A. (2012). Finance-based scheduling for multiple projects with multimode activities. In *Construction Research Congress 2012* (pp. 386–396), May 21–23, 2012.

Fathi, H., & Afshar, A. (2008). Multiple resource constraint time–cost–resource optimization using genetic algorithm. In *First International Conference on Construction in Developing Countries (ICCIDC–I), "Advancing and Integrating Construction Education, Research & Practice", Karachi, Pakistan, August 4–5, 2008.*

Ferrada, X., Serpella, A. F., & Skiniewski, M. (2013, December). Selection of construction methods: A knowledge—based approach. *The Scientific World Journal* (*Impact Factor: 1.73*).

Frost, R. (1986). *Introduction to knowledge-based systems*. Indianapolis: Macmillan Publishing Co., Inc.

Haidar, A. D. (1996, February). Ph.D., Thesis entitled 'equipment selection in opencast mining using a hybrid knowledge base system and genetic algorithms'. London, UK: School of Construction, South Bank University.

Hegazy, T. (1999, May/June). *Optimization of resource allocation and levelling using genetic algorithms*. http://kisi.deu.edu.tr/guzin.kavrukkoca/BISMaster/ResourceLevelingUsingGenetic Algorithms.pdf.

Iranagh, M. A., & Sonmez, R. (2012, September 3–5). *A genetic algorithm for resource levelling of construction projects*. http://www.arcom.ac.uk/-docs/proceedings/ar2012-1047-1054_ Iranagh_Sonmez.pdf.

Maged, G., & Basily Sameh, Y. (2008). Using genetic algorithms in optimizing construction material delivery (2008) Using genetic algorithms in optimizing construction material delivery schedules. *Construction Innovation, 8*(1), 23–45.

Muñoz-Avila, H., Gupta, K., Aha, D. W., & Nau, D. S. (1987). *Knowledge-based project planning*. New York: Springer. http://www.cse.lehigh.edu/~munoz/Publications/Munoz1. PDF.

Sriprasert, E., & Dawood, N. N. (2003). Genetic algorithms for multi-constraint scheduling: An application for the construction industry. In R. Amor (Ed.), *Proceedings of the CIB W78 20th International Conference on Construction IT, Construction IT Bridging the Distance, CIB Report 284, Waiheke Island, New Zealand* (pp. 341–353), April 23–25, 2003.

Winston, P. H. (1979). *Artificial intelligence: An mit perspective: Expert problem solving, natural language understanding and intelligent Computer coaches, representation and learning (Vol. 1)*. Cambridge: MIT Press, Published March 20th 1979.

Yazdani, M. (1986). *Artificial intelligence: Principles and applications*. London: Chapman & Hall Computing Series, Published: Cengage Learning EMEA, June 19, 1986.

www.ingramcontent.com/pod-product-compliance
Lightning Source LLC
Chambersburg PA
CBHW071446230926
51846CB00001B/363
*9783319356488*